Surface Engineering

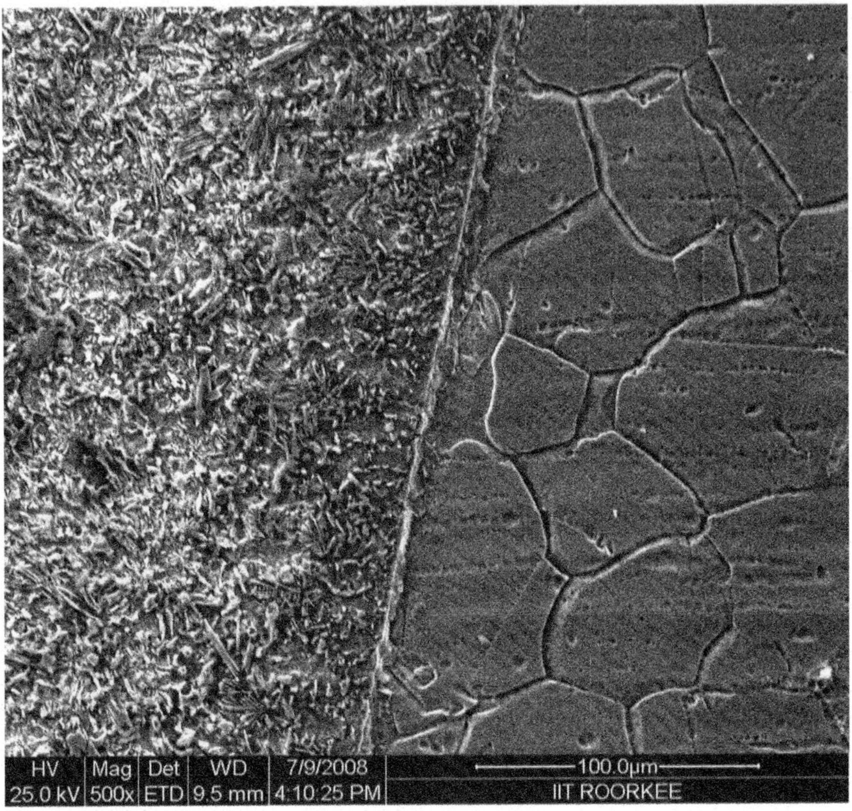
HV
25.0 kV
Mag
500x
Det
ETD
WD
9.5 mm
7/9/2008
4:10:25 PM
100.0µm
IIT ROORKEE

Dheerendra Kumar Dwivedi

Surface Engineering

Enhancing Life of Tribological Components

Springer

Dheerendra Kumar Dwivedi
Department of Mechanical and Industrial Engineering
Indian Institute of Technology Roorkee
Roorkee, Uttarakhand
India

ISBN 978-81-322-3935-2 ISBN 978-81-322-3779-2 (eBook)
https://doi.org/10.1007/978-81-322-3779-2

Softcover re-print of the Hardcover 1st edition 2018

Printed on acid-free paper

This Springer imprint is published by Springer Nature
The registered company is Springer (India) Pvt. Ltd.
The registered company address is: 7th Floor, Vijaya Building, 17 Barakhamba Road, New Delhi 110 001, India

Preface

The wear and corrosion predominantly cause premature failure and reduce the performance of engineering components during service. A new field "surface engineering" is being developed through extensive research across the globe to tackle the wear and corrosion problems. This book provides engineers with enhanced capability to analyze service condition and potential mechanisms of surface degradation so as to select the suitable material for longer service life and improved performance of component and modify surfaces using an appropriate method of surface engineering. The textbook provides matter in lucid and structured manner in seven chapters. This book is well illustrated with schematics, photographs, microstructure, XRD patterns, EDAX mapping, and technical data tables. This book focuses on the influence of materials and methods of surface engineering on structure, properties, and wear performance during the service. This book begins with the need to study the subject of surface engineering, scope of surface engineering, classification of techniques of surface engineering of commercial and academic importance. This book includes conventional material system (steel, cast iron, stellite, WC–Co, PCDs, etc.) and newer materials like multilayer structures, functionally gradient materials (FGMs), intermetallic barrier coatings, and thermal barrier coating. Most of the conventional as well as advanced surface modification techniques such as burnishing, shot peening, flame and induction hardening, laser and electron beam hardening, plasma and TIG melting, carburizing, nitriding, cyaniding, boronizing, ion implantation, laser alloying, chemical vapor deposition, PE chemical vapor deposition, physical vapor deposition, weld overlays, laser cladding, electroplating, electroless plating are included. This book also includes introduction of friction stir processing of aluminum alloys. Further, finding of studies on structure, mechanical and wear properties of weld surfacing, flame spray coating, HVOF-spray coating, laser cladding of ferrous metals, nickel- and cobalt base alloys and their composites in as-spray conditions is included. This book provides an overview of various destructive and nondestructive techniques used for (thickness, soundness, and chemical composition; metallurgical, mechanical, and tribological properties) the characterization of engineered surfaces. This book is expected to be a useful text matter for undergraduate and postgraduate students besides practicing engineering and those interested to pursue the research in the area of surface engineering.

Roorkee, India Dheerendra Kumar Dwivedi

Contents

About the Author

Dheerendra Kumar Dwivedi, Ph.D. is Professor in the Department of Mechanical and Industrial Engineering, Indian Institute of Technology Roorkee. He has been involved in teaching, research and development, industrial consultancy for the last 20 years broadly in the area of manufacturing technologies in general and casting, welding, and surfacing modification for improved mechanical properties in particular. He has developed more than 100 hours of video lectures in the area of manufacturing technologies under NPTEL program of MHRD, Government of India. He supervised 10 Ph.D. theses and more than 50 M.Tech. dissertations. He has published more than 100 research papers in peer-reviewed SCI and SCIE-indexed international journal with h factor 23 and i-10 index 47. He has published a book on production and properties of cast Al–Si alloys with New Age International, New Delhi. He has executed more than 16 R&D projects and 50 industrial consultancy projects of worth Rs. 50 million. He has undertaken five bilateral international collaborative research projects with reputed university, namely Chemnitz University, Germany; University of Coimbra, Portugal; University of Uberlandia, Brazil; University of Zacatecas, Mexico; and Physical Technical Institute, Minsk, Belarus. He has undertaken research projects in the area of oxy-fuel flame and high-velocity oxy-fuel spraying for improved abrasive and erosive wear resistance, laser cladding for improved cavitation resistance, laser-assisted nitriding and ion implantation cast martensitic stainless steel for improved erosion resistance, friction stir processing of cast Al–Si alloys, Ni–Al–bronzes from funding agencies, namely DST, BRNS, MHRD, CSIR. For further details, please visit http://www.iitr.ac.in/~ME/dkd04fme.

Chapter 1
Surface Engineering

1.1 Introduction

The performance of engineering components during service is largely governed by the characteristics of surface and near-surface layers apart from their load-carrying capabilities and service conditions, as the failure in performance mostly begins from the surface or near-surface regions. Therefore, a field of "surface engineering" has been developed, which is primarily related with science and technology of modifying surface characteristics as per needs, so as to enhance the life of the surface engineered product. The need to achieve the desired characteristics at the surface and subsurface regions of an engineering component is primarily dictated by the service conditions and functionality of the component.

1.2 Purpose and Need of Surface Engineering in Industries

The surface engineering is a generic approach toward the enhancement of the performance of engineering component with respect to the following:

- Increase the resistance to corrosion, wear, oxidation, and sulfidation.
- Enhance the mechanical properties, electrical and electronic properties, thermal conductivity, and insulation.
- Reduce the friction coefficient and improve lubrication characteristics.
- Improve aesthetics characteristics.

Further, attempts must be made to reduce the adverse effects of thermal and mechanical loading (if any) during the application of approaches related to surface engineering for modification of the surface properties in relation to the bulk

D. K. Dwivedi, *Surface Engineering*,
https://doi.org/10.1007/978-81-322-3779-2_1

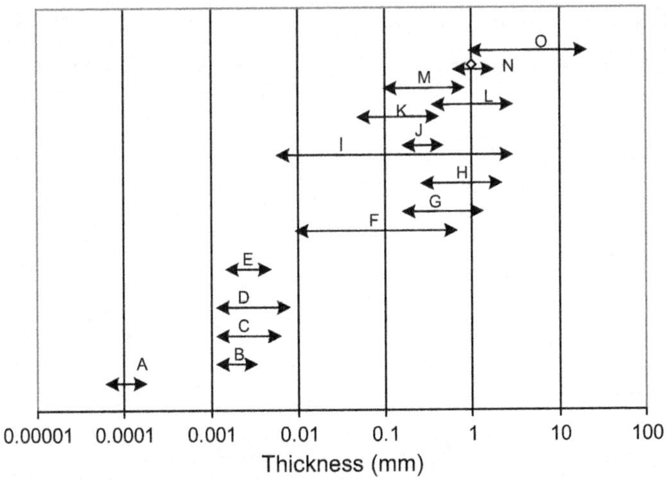

Fig. 1.1 Schematic showing the capability of different surface modification techniques with respect to range of modified thickness or subsurface depth (where A: ion implantation, B: PVD, C: CVD, D: electrolytic plating, E: electroless plating, F: hot dipping (galvanizing/aluminizing), G: laser surface alloying, H: transformation hardening, I: mechanical working, J: nitriding, K: carbonitriding, L: carburizing, M: thermal spraying, N: friction surfacing, O: weld overlays)

material. The design of the engineering components is based on the characteristics of bulk materials used for the manufacturing of the component, so that while operating, the engineering component is able to take main mechanical load. Thus, bulk material and modified surfaces are expected to take service load. The deterioration in properties of bulk material is known to take place if there is prolonged exposure to high temperature. For example, unnecessarily long exposure of components at high temperature during carburizing leads to the coarsening of grain structure of the steel components, which in turn will lead to the deterioration of toughness. Engineered surfaces are expected to offer distinct properties at the surface and near-surface layer only (ranging from few nanometers to tens of millimeters) depending upon the techniques being used (Fig. 1.1).

1.3 Surface and Subsurface Regions

Surface engineering is defined as an approach of developing the surfaces having features different from that of bulk material for enhancing the life and functionality of the engineering component while providing service.

The surface of any component made of crystalline materials is characterized by (a) the nature of surface irregularity which is quantified in terms of surface

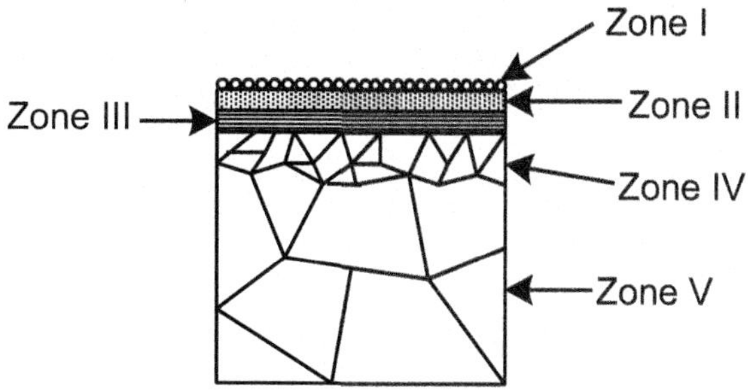

Fig. 1.2 Schematic of different zones commonly observed at surface

roughness and (b) the subsurface region which is generally composed of the following five distinct zones as shown schematically in Fig. 1.2:

- Zone I: Comprises a very thin layer of few nanometers thickness called contamination layer which retains absorbed gases, hydrocarbons, moisture, etc.
- Zone II: Constitutes impurities such as oxides, nitrides, which are formed as a result of interactions between atmospheric or ambient gases and substrate surface.
- Zone III: Involves a work-hardened layer usually thinner than 1 μm with badly damaged crystalline structure.
- Zone IV: Consists of layer of thickness ranging from few microns to hundreds of micrometers with grain structure deformed by the application of external stresses during manufacturing or development of residual stresses.
- Zone V: Involves normal structure of bulk materials as per thermal and mechanical stresses experienced by the material during manufacturing.

1.4 Properties for the Enhanced Life and Performance of Mechanical Components

Tribology performance governing the life of components subjected to wear is dictated by various physical, mechanical, chemical, and dimensional properties of the surface materials of the engineering component in question. The details about the properties of importance for longer tribological life of components can be listed in the following groups:

- **Physical properties**: thermal conductivity, thermal insulation, thermal expansion coefficient, refractoriness, density, optical properties (absorption or reflection of radiations);
- **Chemical properties**: chemical affinity, corrosion, oxidation;
- **Mechanical properties**: hardness, strength, ductility, fracture toughness, bond strength, residual stress, stress corrosion cracking, hydrogen embrittlement;

- **Dimensional properties**: straightness, flatness, roundness, surface roughness, length, etc.

The surface properties required for any component are solely determined by its application and purpose (service conditions), which is ascertained based on the combination of properties necessary for building the component. For example, a combination of toughness and hardness, corrosion resistance, and smoothness will be important for developing pump parts that have good cavitations resistance in association with the enhancement of the thermal efficiency of heat exchanger for augmenting thermal insulation using thermal barrier metal systems at the surface. The significance of specific properties on material removal by different wear mechanisms will be elaborated in the subsequent chapters and sections.

1.5 Classification of Surface Modification Techniques

Surface modification processes primarily involve the following two approaches: (a) bringing change in one or more zones of subsurface as described in Sect. 1.2 and (b) developing another layer of suitable material at the surface to achieve the properties desired for the enhancement of tribological life of the component. In the former case, the changes at the surface and subsurface zone are primarily made in two ways: (a) changing the structure of surface layers using thermal and mechanical methods without making any compositional modification and (b) changing the chemical composition (chemistry) of surface and subsurface layers. In the case of latter, a layer of suitable material (similar or different from the substrate) is developed on to the surface of component for realizing the desired surface properties. In view of the above, the surface modification techniques can be classified as follows.

1.5.1 Changing the Structure of Surface and Near-Surface Layers

a. Mechanical method: Based on localized plastic deformation and so as to achieve work hardening of near-surface layers

    i. Burnishing
    ii. Shot peening

b. Thermal methods: Based on localized heating and controlled cooling to obtain desired microstructure

    i. Flame and induction hardening
    ii. Laser and electron beam hardening
    iii. Plasma and TIG melting

1.5.2 Changing the Chemical Composition of Surface and Near-Surface Layers

c. Diffusion-based processes

   i. Carburizing: Introducing carbon in low carbon steel;
   ii. Nitriding: Introducing nitrogen in ferritic steel;
   iii. Cyaniding: Increasing concentration of both carbon and nitrogen in steel;
   iv. Boronizing: Introducing boron in steel;
   v. Vanadizing: Introducing vanadium in steel.

d. Ion implantation: Introducing nitrogen and other elements and controlled lattice deformation
e. Laser alloying: Based on the intermixing of alloying elements in the substrate using controlled melting of near surface layers

1.5.3 Developing a Surface Layer or Overlays

f. Diffusion-based processes

   i. Chemical vapor deposition
   ii. Physical vapor deposition

g. Melting-based methods

   i. Weld overlays
   ii. Laser cladding

h. Dipping in hot melt-based methods

   i. Hot dip galvanizing
   ii. Hot dip lead-tin coating
   iii. Hot dip aluminizing
   iv. Hot dip chromizing

i. Electrolysis-based methods

   i. Electroplating of

      1. Cr and Ni for esthetics;
      2. Cd and Zn for controlling corrosion;
      3. Cu and Ag for improving electrical properties;
      4. Hard chromium for enhancing wear resistance.

   ii. Electroless plating (Ni–P and Ni–B) for improved corrosion and wear resistance

j. Mechanical methods

   i. Mechanical plating
   ii. Roll bonding
   iii. Explosive bonding
   iv. Hot isostatic pressing

Different processes can be compared with respect to the following parameters. Comparison of few important surface modification techniques with respect to the technology level and complexity is schematically shown in Fig. 1.3.

1. Capability

   a. To handle the material of low or high melting points;
   b. To modify components up to certain size, area, thickness, and depth;
   c. To apply surface modification under fabrication constraints at site or shop;
   d. To reduce thermal or mechanical or tribological or chemical damage on the surface of workpiece;
   e. To produce smooth surface, Ra;
   f. To provide control over the surface modification processes.

2. Initial investment, availability, and expertise needed.

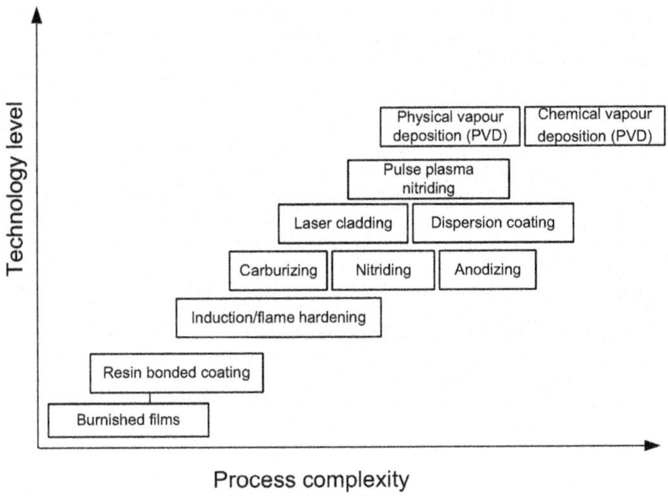

Fig. 1.3 Schematic comparison of various surface modification techniques based on process complexity and technology level

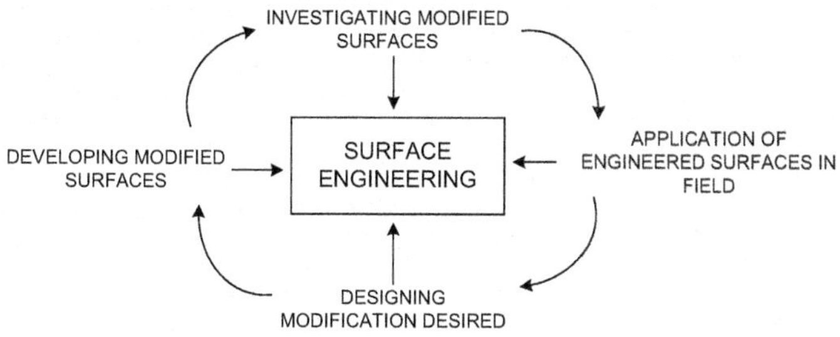

Fig. 1.4 Block diagram showing scope of surface engineering

1.6 Scope of Surface Engineering

The scope of surface engineering basically involves four broad aspects: (a) designing the surface to be developed, (b) developing or modifying the substrate using suitable approach, (c) characterizing the engineered or modified surfaces, and (d) putting them in application for the purpose (Fig. 1.4).

1.6.1 Designing Surface Modification

Designing of surface modification primarily involves identification of the type and characteristics of surface (to be achieved in light of service condition), which is crucial for the successful application of the engineering component while offering service. Surface properties desired for the fulfillment of the service life required are determined by the service conditions with respect to load, environment, and reliability. The application of selected surface modification process on to the substrate helps in achieving the designed surface characteristics. Tribological components are subjected to different environments and stress conditions; hence, there is the requirement for these components to possess varying set of properties for longer life. These properties are corrosion resistance for those working in moist environment, hardness for low stress abrasion condition, fracture toughness, and fatigue resistance for fluctuating load condition such as gear tooth surfaces. For the designing of surface modification in relation to a component for a given service, the compatibility of materials and approach with substrate material, configuration, and location must be kept in mind.

1.6.2 Developing the Modified Surface

In view of design characteristics, the substrate surface in question is engineered or modified using material, method, and procedure established or identified at the design stage. Further, the relevant procedure for modification of the surface must be investigated and optimized by conducting pilot studies. After establishing the

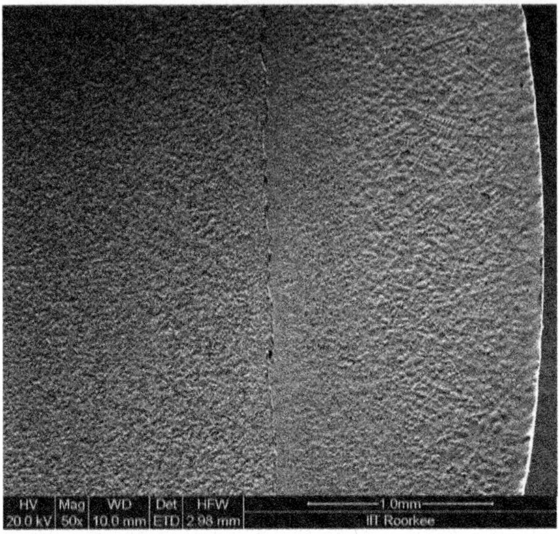

Fig. 1.5 Scanning electron macrograph of the flame sprayed coating with coating on the right, substrate on the left, and interface in the middle

procedure for surface modification based on the requisite designed properties, the surfaces of the real component are modified. A macrograph of a typical coating developed on the structural steel substrate after optimization of the process parameters by flame spraying is shown in Fig. 1.5. The principles and procedures for developing modified surfaces are specific to the utilized techniques. As per the selection of the surface modification process, these approaches can vary appreciably from mechanical to thermal and to electrolytic approaches.

1.6.3 Investigating and Characterizing the Modified Surfaces

Modified surfaces are examined for evaluation of the parameters affecting the quality or performance of the modified surface while offering service through the utilization of suitable destructive and nondestructive techniques. For example, hardness of the coating shows resistance to abrasion and indentation (Fig. 1.6). The size of indentation under identical loading condition is used as a relative measure of hardness. The larger the size of indentation, the lower the hardness. Characterization of the modified surfaces is done for three purposes: (a) to assess the soundness of modified surfaces, that is, the presence of discontinuity such as porosity, bonding, inclusion, and cracks, if any, (b) to quantify the properties of modified surfaces affecting their performance with respect to the chemical composition, microstructure, hardness, bond strength, etc., and (c) to measure the performance of modified surfaces largely under simulated conditions in correspondence to real life, for example, adhesive, abrasive, erosion, fretting, corrosion resistance, optical and electrical properties.

Fig. 1.6 Scanning electron micrograph showing indentations developed during the hardness test of the coating and coating–substrate interface

1.6.4 Application of Modified Surfaces

The engineered surfaces are then put in use for offering service, and their performance is evaluated generally based on the ratio between the component's life (in terms of time or number of units manufactured or quantity handled) without surface modification and the component's life after surface modification under identical service conditions. This ratio indicates the effectiveness of the surface modification of the substrate. The less the value of the ratio, the greater the improvement in the tribological life of components after surface modification. In case of any deviation from the intended performance of the modified surface, it is required to repeat stages of designing surface modification followed by developing modified surfaces, investigating and characterizing the modified surfaces and application of the same on the engineering components.

1.7 Role of Surface Properties Affecting Wear and Friction Behavior

There are five characteristics of the surfaces, namely **surface energy**, **chemical composition and microstructure**, **hardness**, **and surface roughness** which affect the wear and friction behavior significantly. However, all these properties except the surface roughness are interdependent and are influenced by thermal and mechanical history experienced by the surface of substrate during the service. A variation in chemical composition generally changes the microstructure, hardness, and surface energy. Surface roughness is determined by the method of manufacturing used for the generation of the surface which is invariably characterized in terms of Ra (centerline average value of height/depth of peaks and valleys present at the surface).

1.7.1 Surface Energy

Surface energy (γ: J/m^{-2}) refers to the energy which is required for creating a unit area of the new surface after breaking the cohesive bond of a material. Therefore, surface energy is primarily determined by the type of bond (covalent, ionic, metallic, van der Waals, hydrogen or molecular bond, etc.) between atoms. The surface energy of materials forming covalent bond (1000–3000 mJ/m^{-2}) is generally higher than the ionic bond (100–500 mJ/m^{-2}) and molecular bond (<100 mJ/m^{-2}). Surface energy decreases with the increase in temperature (owing to increase in the amplitude vibrating atoms). Further, the adsorption of impurities at the surface level of a material can lead to appreciably decrease in the surface energy (ten to hundred times) of the material.

The surface energy of a material is measured with respect to the surrounding environment forming an interface at the surface: (a) solid–liquid interface (γ_{SL}), (b) solid–vapor interface (γ_{SV}), and (c) liquid–vapor interface (γ_{LV}).

Moreover, the surface energy of solid–solid interface (γ_{SS}) for an object determines the fracture resistance, while the surface energy of an object with liquid–vapor interface influences the friction and wear behavior. In general, low surface energy of an object with solid–vapor interface (γ_{SV}) results in less wear and hence acts as an excellent wear resistant. The surface energy with solid–vapor interface (γ_{SV}) in turn depends on the dihedral angle and grain boundary energy, γ_{gb} (Fig.1.7).

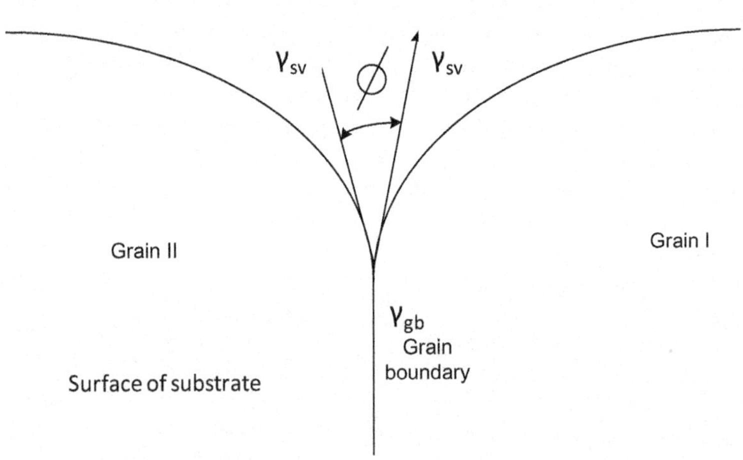

Fig. 1.7 Schematic of grain boundary energy and solid vapour surface energy relationship

$$\gamma_{gb} = 2.\gamma_{SV} \times \cos(\text{dihedral angle}/2)$$

In general, low surface energy (γ_{SV}), high dihedral angle, and few grain boundary precipitations result in improved wear resistance. Fine grain structure produces large grain boundary area which results in high total grain boundary energy at the surface. Hence, small grains produce better resistance toward yielding and fracture.

According to Alison and Rabinowitz, surface energy (γ) and surface energy at solid–vapor interface (γ_{SV}) are directly (related) proportional to friction coefficient and wear behavior, respectively. In addition, stacking fault energy is found to be directly proportional to wear loss and inversely proportional to work hardening. A low stacking fault energy causes more twins on plastic deformation which in turn offers increased interfering effect on slip that leads to greater resistance to deformation. Stacking fault energy being inversely proportional to work hardening. Lower the energy, greater the work hardening which leads to reduction in wear and enhancement in wear resistance.

The surface energy can also be understood as the energy of surface that is in equilibrium with the surrounding. This comprises energies resulting because of grain boundaries, twin boundaries, and energies associated with locked-in strains owing to non-uniform plastic deformation, thermal cyclic heating, and related transformations.

1.7.2 Surface Composition

The chemical composition of the surface predominantly determines the wear behavior, as it directly affects the surface energy, microstructure, and hardness of the component. The change in composition of surface affects the phase and grain structure of microconstituents such as matrix, precipitates, and other intermetallic compounds present at the surface, which in turn determine the hardness, surface energy, and work hardening tendency. Further, variation in the surface composition may cause grain refinement, solid solution (interstitial or substitutional) hardening, precipitation, and dispersion hardening, and change in stacking fault energy (affecting the work hardening behavior), which in turn can appreciably affect the wear resistance owing to variation in hardness, surface energy, and work hardening tendency.

The surface composition can be modified using thermal energy based on two approaches: (a) changing the surface chemistry on the basis of diffusion principle and (b) developing the surface with the material having the desired composition for the fulfillment of the properties required. Thermal energy-based approaches help either in the diffusion of alloying element or in controlled melting of the substrate to introduce the element and compound into the surface for the fulfillment of the required modification in the surface properties.

In case of the first approach, the chemistry of surface is primarily changed through the diffusion of required elements such as C, N, Cr, Mo, and B. The surface of the component to be modified is exposed to the environment rich in the required element (solid, liquid, or gaseous form) at a high temperature for a long time. The presence of these elements either singly or in combination with other elements can lead to formation of solid solution and intermetallic compounds besides assisting in hardening through phase transformation to improve the surface characteristics.

In case of the second approach, a coating or layer of another material (compatible with the base metal) which can provide the desired combination of properties is developed using suitable approaches:

A. Thermal spraying: Chemistry of the base metal is not affected.
B. Spray and fuse: Some change in the chemistry of base metal takes place through diffusion.
C. Welding: Melting of the substrate changes the chemistry of overlaying material due to dilution.
D. Electroplating and dipping in molten bath: No change in chemistry of the substrate.

1.7.3 Surface Microstructure

The microstructure of any material includes two aspects: (a) phase structure indicating different types of phases, relative amounts of phases present, and their distribution in the matrix and (b) grain structure showing size and shape of different microconstituents and their distribution. Any change in one or more of the above six parameters of the microstructure brought in by the mechanical and thermal processing and surface chemistry modification can influence the wear and corrosion behavior due to the variation in electrochemical and mechanical properties.

Phase structure
Surface modification through change in the chemical composition results in the formation of either hard phases (such as carbide, borides, nitrides) or the phases having low stacking fault energy (such as austenite in high Mn–Ni steel). The phases with low stacking fault energy tend to work harden rapidly, which in turn increases the wear resistance especially under the impact of load conditions during service. Moreover, the formation of discrete hard microconstituents (CrC, TiB, WC, VC, TiN, TiC) in the matrix also increases the wear resistance. The presence of the discrete hard microconstituents provides better support to the matrix for taking up the load and gives the ability to obstruct the path of abrasion and scratches during wear. These factors in turn reduce wear.

The relative amount of hard, soft, and work hardening phases and precipitates in the microstructure of engineered surfaces affects the wear and friction behavior. However, the extent of influence that the relative amount of various phases has on the wear and friction behavior depends on the impact that the microconstituents have on the hardness, work hardening behavior, and surface energy.

Further, the non-uniform distribution of the phases in the modified surface adversely affects the performance of the modified surfaces. For example, the presence of excessive amount of hard phases such as iron nitrides at the surface of steel after nitriding leads to crack and makes the surface layers fragile. Moreover, the loss of certain alloying elements from the near-surface layers of coatings developed by the thermal spraying through oxidation reduces the tribological performance of engineered surfaces under the adhesive and abrasive wear conditions. Similarly, the deterioration in the quality of surfaces modified by welding is also observed due to dilution (intermixing of molten substrate and filler metal).

Grain structure

The grain structure involving the size and shape of microconstituents present at the surface affects the surface energy, yield strength and hardness, and cracking tendency at the particle–matrix interface. According to Hall–Patch relationship, a decrease in the grain size generally increases the resistance to movement of dislocations over long distances, which in turn increases the resistance to plastic deformation (yield strength) and hardness (Fig. 1.8a). The size of microconstituents can vary significantly from submicron size to few thousands of microns. The grain sizes corresponding to the different ASTM numbers are shown in Fig. 1.8b. Further, the fine grain structures require higher stress for the nucleation of cracks and voids at the particle–matrix interfaces, thereby demonstrating that grain refinement improves the fracture toughness and wear resistance. The influence of grain size on surface energy has been elaborated above.

The shape of microconstituents can vary significantly from columnar, cellular, dendritic, equiaxed, needle, acicular, nodules, spherical, cuboid and polyhedral, star

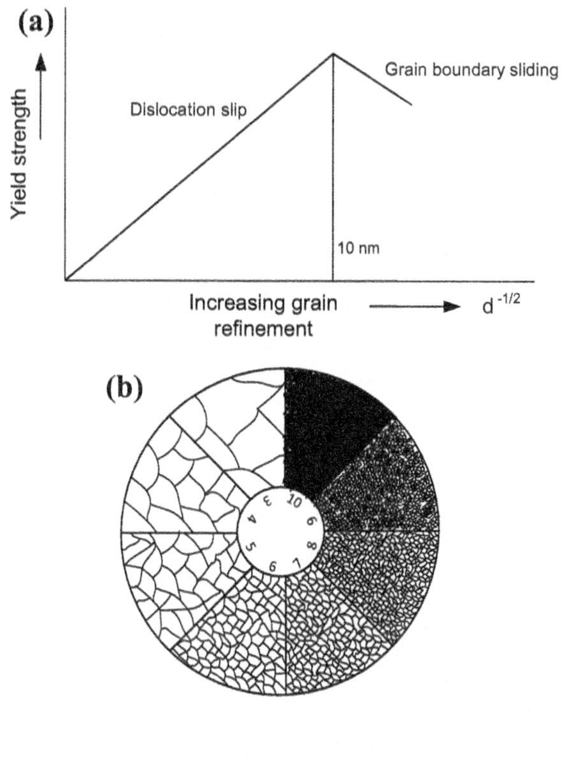

Fig. 1.8 Schematic showing **a** the effect of grain size on yield strength with Hall–Patch strengthening limit and **b** the relative grain sizes as per ASTM number

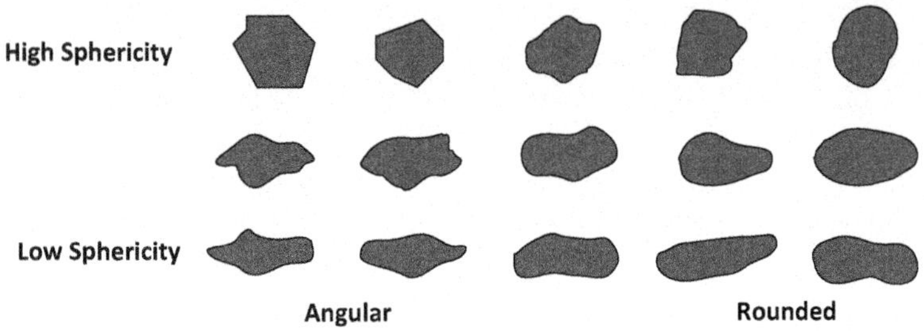

Fig. 1.9 Schematic showing of different grain shapes

and Chinese script, etc. The morphologies of microconstituents influence wear depending on the ways these affect the crack nucleation tendency and their growth at the particle–matrix interfaces. In general, needle-shaped sharp-tipped acicular grain morphologies are found to be more prone to cracking and void formation than the spherical and round-tipped microconstituents. Therefore, the modified surfaces having needle-shaped sharp-tipped hard microconstituents show higher wear rate than the other morphologies. Form factor is defined as the ratio of length to width of microconstituents, which is a commonly used parameter for characterizing the shape or morphology of the microconstituents (Fig. 1.9).

1.7.4 Surface Roughness

The roughness of functional surfaces of tribological components affects the friction and wear behavior. Surface roughness indicates the extent of peaks and valleys present on the mating surfaces. The increase in microlevel ups and downs on the surface in the form of peak and valleys enhances the mechanical interlocking and cold bond tendencies between the asperities present on the mating surfaces, which in turn promotes the frictional resistance and the tendency of removing the material from the interacting surfaces in the form of debris under adhesive wear conditions. Asperities on the hard surfaces act as indenter that results in developing scratches on the soft surfaces due to the relative movement between the mating surfaces under abrasive wear conditions. Under erosive wear (jet, solid particle and cavitation), surface fatigue, stress corrosion, and fretting wear conditions, deep valleys present on the surface act as stress raisers which facilitate the formation of wear debris by easy nucleation and growth of cracks. In general, high surface roughness increases the wear rate of metallic components. It is worth noting that during the interaction between the mating components, surface roughness continuously changes, but it stabilizes in the long run unless there is a change in wear mechanism owing to variation in service conditions such as load, velocity, lubrication, presence of foreign particles. Further, surface roughness affects the size of wear particles that are produced under adhesive and abrasive wear conditions. In general, increase in surface roughness enhances the size of wear particles produced during the interaction of mating surfaces which in turn enhances the wear rate. The equilibrium peak-to-peak roughness values have good correlation with the mean wear particle size generated during the interaction between same material tribological systems.

1.8 Advantages, Limitations, and Applications

Surface modification can be applied in both old or used component as well as newly manufactured component for increasing the service life and functional capability of working under more severe and hardworking condition even while using the cost-effective and somewhat low-quality substrate material. However, the substrate should have enough mechanical and other relevant properties in order to fulfill the service load conditions because surface modification helps in improving the surface properties and functionalities. The advantages of surface engineering can be enlisted as follows:

- Increases the tribological life of component by reducing wear rates;
- Increases reliability and mean time between repairs;
- Increases the capability of withstanding at elevated temperatures by using thermal barrier coating;
- Reduces the cost of item for a given service life;
- Increases the resistance to stress corrosion cracking, fatigue, and tensile stresses by developing residual compressive stresses.

Issues with inappropriate surface modification

Inappropriate surface modification of substrate can lead to various issues (a) the integrity of developed surface, particularly with respect to the cracks, pores, inclusion in the forms of oxides, and slag specially in those surface modification processes which involve melting and solidification, (b) the anisotropic behavior of coating related to the surface and coating–substrate interface or substrate itself, (c) the strength of bonding between surface layers or coatings which do not involve the melting of substrate and diffusion for surface modification, and (d) the physical, chemical, mechanical, and metallurgical incompatibilities between the developed surfaces and substrate may lead to residual stress, embrittlement of interfaces, degradation in properties of surfacing, and peeling off or delamination of coating from the substrate (Fig. 1.10).

Excessive nitrogen and carbon concentration in the near-surface layers cause embrittlement which results from nitriding and carburizing of steel, and this is a typical example of anisotropic characteristics of modified surfaces.

Applications of surface engineering

- Increase the resistance toward abrasion in mining, petrochemical industry in production, and exploration using Fe–Cr–C, Ni–WC, Co–Cr–W–C system.

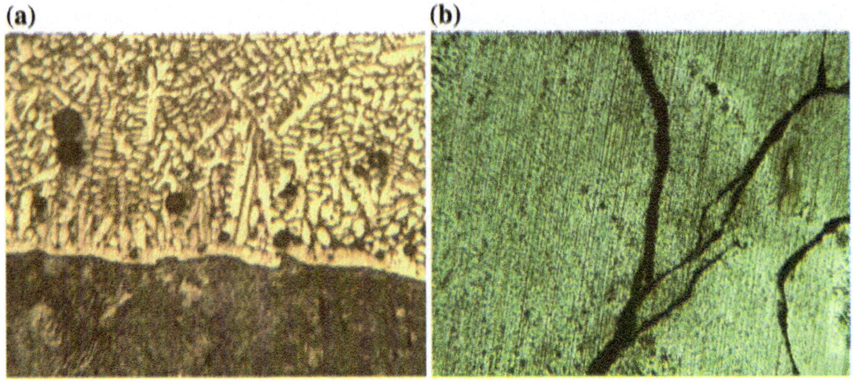

Fig. 1.10 Typical micrographs of coatings showing discontinuities such as **a** porosity and **b** cracks

- Increase the cavitation and erosion resistance in hydro-turbines using Fe–Mn–C, Fe–Cr–C, Co–Cr–W–C systems.
- Increase the corrosion resistance of steel substrate in general engineering industry using stainless steel, Zn and chromium plating.
- Increase adhesive wear resistance using chromizing.
- Improve the optical and electrical properties using Au and Ag coating.
- Increase the capability to withstand the elevated temperature using thermal barrier coating.
- Reduce friction between mating component using self-lubricating materials such as graphite, MoS_2.
- Enhance the mechanical performance of substrate by coating or surfacing using suitable materials.

Questions for self-assessment

1. Describe the factors affecting the wear and friction behavior of tribological components.
2. Give the classification of surface modification techniques.
3. What is the scope of surface engineering?
4. Explain the different zones near the surface layers using schematic diagram.
5. How is surface engineering useful for enhancing the performance?
6. Will there be difference in the tribological performance of component of the given composition made by different manufacturing processes, namely casting and rolling?
7. How do surface irregularities affect the adhesive wear resistance?
8. What is the importance of proper designing of surface modification?

Chapter 2
Surface Damage: Causes and Mechanisms

2.1 Material Properties and Its Effect on Performance of Components

The performance of engineering component during service is predominantly determined by the surface characteristics because their failure is mostly triggered from the surface. Degradation at the surface of the component owing to any physical, chemical, and biological phenomenon/activities leads to loss of materials (wear), corrosion, and fracture. These degradations do not just lead to the loss of materials from the surface but also deteriorate the surface finishing by developing severe undulations at the surface through the formation of oxides and other compound based on the interactions with the surrounding environment. The undesirable features of surface degradation can degrade the performance of engineering components below the critical level, thereby forcing the requirement of repairing or replacing the engineering component with new ones (Fig. 2.1). Surface engineers are, therefore, focusing on the development of technologies for reducing surface degradation rate so as to increase the service time by improving surface qualities, to enhance the performance and life of components.

2.2 Common Factors Leading to the Deterioration of Surfaces

The deterioration of functional surfaces begins as soon as they are put in use owing to the interaction of surfaces with heat, radiation, mechanical stresses, ambient and exhaust gases, water and other liquids, biological items, and their wastes during the service. On the basis of fundamental mechanism, various factors causing the deterioration of functional surfaces (due to interactions during service) can be classified or grouped in three categories as shown schematically in Fig. 2.2. This

D. K. Dwivedi, *Surface Engineering*,
https://doi.org/10.1007/978-81-322-3779-2_2

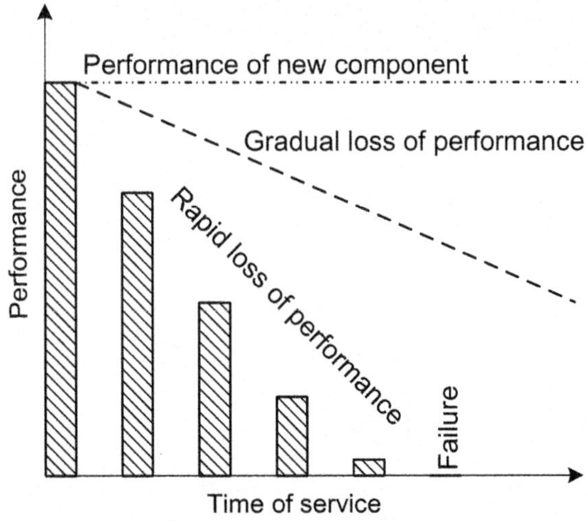

Fig. 2.1 Performance versus time relationship for any engineering component

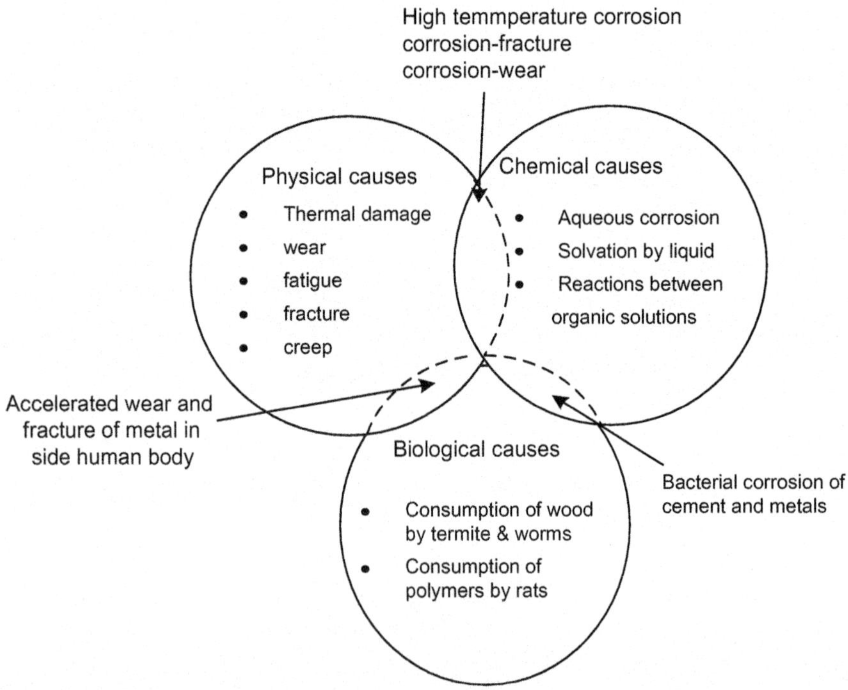

Fig. 2.2 Schematic diagram showing the classification of the causes of surface degradation

kind of classification of the surface degradation mechanisms is primarily for developing strategy to control the surface damage. Surface damage is caused because of the following three reasons: (a) Physical phenomenon is primarily caused by heat, mechanical loads, and radiation, (b) chemical phenomenon is caused due to chemical reactions of surface with water, atmospheric gases, chemicals, and waste items, and (c) biological phenomenon is caused due to interactions of surfaces with biological items and their wastes.

The development of strategy (for surface modification using suitable approach) to control the surface degradation is generally found to be easier when damage occurs due to a single mechanism (such as wear, corrosion) rather than due to simultaneous operations of two or more mechanisms, such as corrosive wear, corrosion fracture, failure of metal. Human body is a typical example of such cases where physical, chemical, and biological phenomena act together.

For controlling the surface degradation under abrasive wear conditions, the surface modification can be designed to have requisite hardness while that for solid particle erosion can be done by developing engineered surfaces having suitable combination of both hardness and toughness. However, abrasion and erosion occurring either in the corrosive environment or at high temperature make the job of surface engineers more complex primarily because of the need of contradicting design property requirement and entirely different nature of operating mechanisms which cause surface damage.

2.3 Types of Wear and Mechanisms and Classical Governing Laws

Wear is one of the most commonly encountered surface degradation mechanism in which gradual and progressive loss of material from the functional surface takes place. This uncontrolled and non-uniform loss of material from the surfaces results in surface undulation and increases surface roughness which eventually leads to change in the size and shape of component to such an extent that it promotes other surface degradation mechanisms, weakens the component and makes it prone to mechanical failure under static or dynamic service loads, and leads to improper movement or malfunction of other mating components or the entire system.

Wear of material from the functional surfaces of engineering components primarily occurs through common mechanisms such as adhesion, abrasion, erosion, corrosion, surface fatigue, fretting. Adhesion, abrasion, and corrosion together contribute to more than 80% of material loss from the engineering components in general. However, the percentage contribution of each wear mechanism varies appreciably with the change of sector such as transport, mining, machinery. In the following sections, the basic principle and factors affecting material loss based on the above wear mechanisms are presented.

2.3.1 Adhesive Wear

Adhesive wear takes place when the interacting surfaces develop a relative motion with respect to each other and are in direct metal-to-metal contact through asperities (peaks and troughs) present at the mating surfaces (Fig. 2.3). These asperities are always present on the mating surfaces irrespective of the manufacturing process used for making the components in question. Even polishing and finishing can just reduce the extent of peaks and valleys present on the surface but cannot eliminate them. In fact, all real contacts occur through asperities at the interacting surfaces; therefore, the actual area of contacts (shaded zones in Fig. 2.3) is just about 0.1–1% of the nominal area of contact, thereby leading to the development of true stress at real contact areas which is approximately 100–1000 times of the nominal stress owing to either the dead weight of interacting member or the external loads. Under such high stress conditions, asperities (peaks and troughs) become plastically deformed and develop direct metal-to-metal contact between the interacting surfaces. This deformation and establishment of direct metallic contact between the mating surfaces result in the formation of cold bond or joints owing to the following three phenomena (Fig. 2.4).

(a) Mechanical interlocking: The interaction of asperities at the mating surfaces forms mechanical joints owing to the interlocking of surface irregularities due to deformation. Moreover, the strength of such cold weld bond at the interface is primarily governed by the work hardening of surface layers owing to deformation, thus making the surface layers harder than the subsurface layers and core materials.
(b) Electron exchange: The development of metallic bond at the interfaces through the exchange of free electrons from one side to other (high energy to low energy side) happens very rapidly when contacts are established just for a fraction of second such as in the case of sliding.

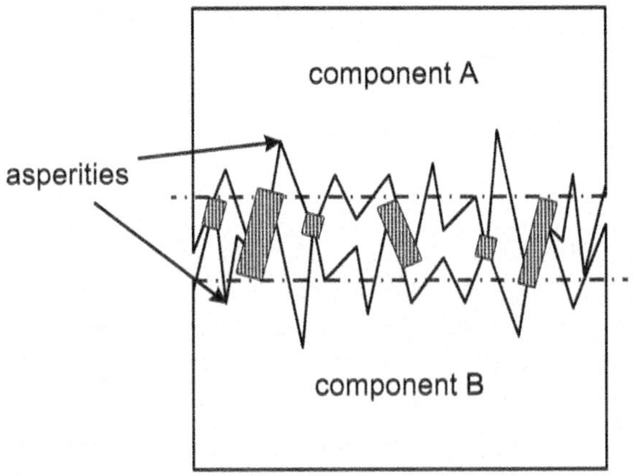

Fig. 2.3 Schematic of contacts at interface in adhesion

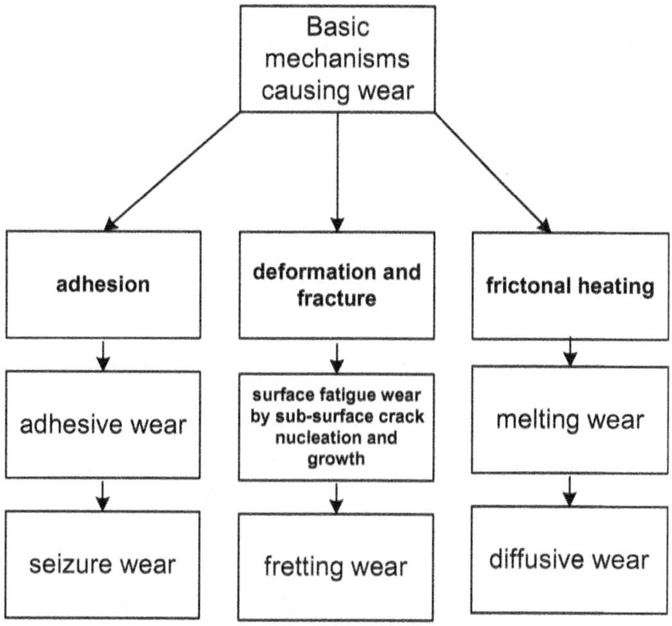

Fig. 2.4 Block diagram showing mechanism of adhesive wear

(c) Diffusion under high temperature at the interface for longer duration leads to the diffusion of alloying elements present in the mating components. Diffusion begins from the areas where direct metal-to-metal contacts exist, and this leads to the development of diffusion at the interfaces.

Bonds formed at the interface do not affect the surfaces of interacting components until the relative movement takes place. For relative movement to happen between the mating surfaces, these bonds must be broken. The fracture of interfacial bond preceding the relative motion can occur from any of the three regions: (a) just at the bond interface, (b) subsurface layer zone of mating one component, or (c) other component whichever is weak. In general, the interface bonds are found stronger than subsurface layers due to the work hardening effect of plastic deformation of the surface layers. Therefore, a small lump of metal is detached generally from the subsurface layers of weaker component of the two interacting members. These particles initially tend to roll and plow at the interacting surfaces due to relative motion. Continued removal of material from the functional surfaces helps these particles to grow. Eventually, these are pushed out from the interface in the form of mostly black oxidized wear debris. The presence of this work hardened piece of metal at the interface can act as the third body for abrasion to take place and increase loss of materials from the surfaces.

The resistance to breaking bonds developed at the interfaces primarily determines the force (friction force) required to have the relative movement. All factors encouraging the formation of strong interfacial bonds between the matting components would lead to increase in the wear and friction effects. There are numerous factors and some of them are: the presence of asperities, load (influencing the true stress at contact area), contact duration, surrounding (atmospheric gases or vacuum), counter surface, metallic intimacy, complexity of microstructure, hardness of mating components, extent of work hardening owing to deformation of near-surface layers, etc.

These factors can be grouped under two headings: (a) service or sliding conditions such as load, relative speed (affecting interface contact duration), interface medium, or air (if any) affecting the metallic intimacy, and (b) material-related parameters such as surface roughness, hardness, ductility, work hardening tendency, complexity of structure, presence of nonmetallic components (such as ceramics, polymers).

In general, increase in normal load (so true stress), surface roughness (so depth and height of troughs and peaks, respectively), contact duration at the interface (at low relative velocity between mating parts), absence of active or oxidative environment, vacuum, inert environment, or absence of lubricant, and high temperature increase the interfacial bonding tendency which in turn enhances the wear and frictional losses.

Sliding of mating component in the ambient conditions causes oxidation of interacting surface layers, thereby reducing the metallic intimacy between them, which in turn decreases the tendency of interfacial bond formation that leads to the reduction in friction coefficient and wear. However, the relative motions occurring between the mating components in vacuum or inert atmosphere do not develop any such protective oxide layers at the surfaces and thus cause direct metal-to-metal contact as well as develop seizure-like conditions.

The interacting surfaces of similar metals having low hardness and high ductility as well as low work hardening tendency further increase the wear and friction effects. Therefore, dissimilar mating metal systems with high hardness and low stacking fault energy help in reducing the wear- and friction-related problems.

Mating components composed of complex microstructures having nonmetal components (like ceramic, carbides, nitrides, and graphite) imbedded in metal matrix (ferrite, pearlite, aluminum, nickel) discourage the interfacial bonding through free electron transfer. As free electron transfer between metal (aluminum, iron, cobalt, etc.) and nonmetal system (silicon, graphite, and ceramic particles) does not occur under normal conditions unless meting takes place. Hence, the reinforcement of hard nonmetallic constituents in metal matrix (as in case of MMCs) generally reduces the friction coefficient and increases the wear resistance.

Mild and severe wear

Under severe sliding conditions of normal load and relative velocity, the generation of excessive frictional heat causes thermal softening and even partial or complete melting of surface layers, which can lead to the situation of (a) 100% direct metal-to-metal contact and seizure-like conditions owing to the thermal softening of surface layers, (b) metallic failure of surface layers due to the inability to take up the service loads, and (c) melting and deposition of partially or melted surface layers on to the counter surface and the generation of wear debris in the form of big metallic particles indicating the occurrence of metallic wear. Metallic wear becomes 10–100 times greater than normal mild oxidative. Tribological components are generally designed to work under mild oxidative wear conditions; however, accidental situations like failure of lubricating, overloading, and overspeeding can lead to metallic wear conditions which will eventually lead to failure of the system.

All the metal systems having high thermal stability and the hard and coherent surface layers offer good resistance to the occurrence of metallic wear. The conditions leading to the occurrence of transition from mild oxidative to severe metallic wear are called transition load or transition velocity.

Factors affecting the adhesive wear

There are many test material, counter surface and sliding test-related parameter, which affect degree of metallic intimacy and mechanism of material loss from the contacting surfaces. The most of sliding adhesive wear factors include normal load, sliding speed, type of relative movement (unidirectional, reciprocating), counter surface (hardness, metal, crystal structure, temperature), environment affecting extent of interfacial contact (ambient air, steam, liquid metal, etc.), nature of contact (pin-on-disc, pin-on-flat, ball-on-flat, ball-on-ring) determining the contact stresses at the interface and material properties (hardness, oxidation tendency, thermal stability, resistance to softening, type of oxide, metallurgical structure, crystal structure), work hardening tendency (work hardening exponent, work hardening capacity, stacking fault energy) affect the wear behaviour.

Normal load: Increase of load during sliding results in higher metallic intimacy under adhesive wear condition which in turn in general increases the wear rate. The rate of change in wear rate due to increase in normal load is determined by operational wear mechanism for a given set of conditions (Fig. 2.5). Increase in wear rate

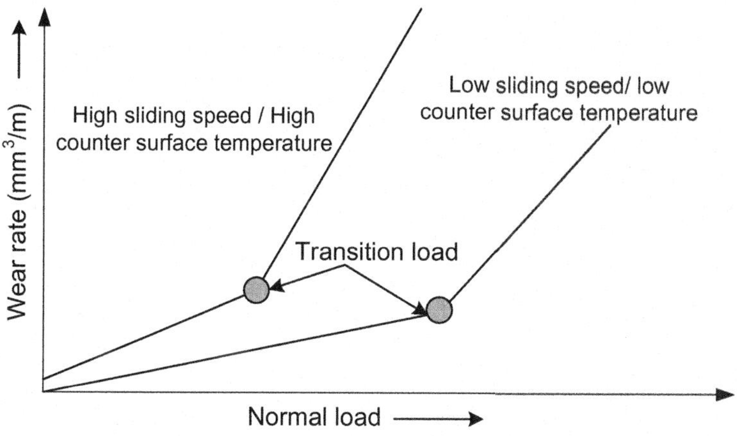

Fig. 2.5 Effect of normal load on wear rate at different sliding speeds and counter surface temperature conditions

as function of normal load is found to be low and linear under mild oxidative wear condition. Under mild wear conditions, wear surface remains coherent and does not experience large-scale plastic deformation. Increase in normal load above a critical value called transition load unstabilizes the wear surfaces causing large-scale plastic deformation and metallic failure. Moreover, the critical load at which change in the wear mechanism from mild oxidative wear to severe metallic wear takes place is influenced by many sliding wear conditions namely sliding speed, counter surface temperature, and its metal (crystal structure, microstructure). In general, high sliding speed, high counter surface temperature, and same crystal structure decrease the transition load. The change in magnitude of wear rate owing to transition from mild to severe wear may be 10–100 folds or even more. Additionally, wear debris changes from dark black oxide particles of fine size (few tens of micrometers) to bright metallic particles of large size (few thousands of micrometers).

Sliding speed: The relative speed between the material of component and counter surface is called sliding speed. Sliding speed affects frictional heat generation and localization of heat generated. Rise in temperature due to frictional heating causes (a) oxidation of interacting surfaces, (b) thermal softening of near sliding surface layers, and (c) complete/partial melting of element at the interface. However, minor temperature rise (about 10–30 °C) occurring under mild wear conditions, i.e., low normal load and low sliding speed, primarily causes oxidation of interacting surfaces. These oxides reduce the metallic intimacy between the mating surfaces which in turn decreases adhesive wear. Therefore, under low-load sliding conditions, initially an increase in sliding speed reduces wear rate. Increase in sliding speed above a certain limit increases in heat localization at mating surfaces to such an extent that thermal softening dominates over oxidation (Fig. 2.6). Increased thermal softening of the sliding surfaces increases the metal-to-metal contact which may lead to (a) increase in wear rate with increase in sliding speed and (b) metallic failure of the

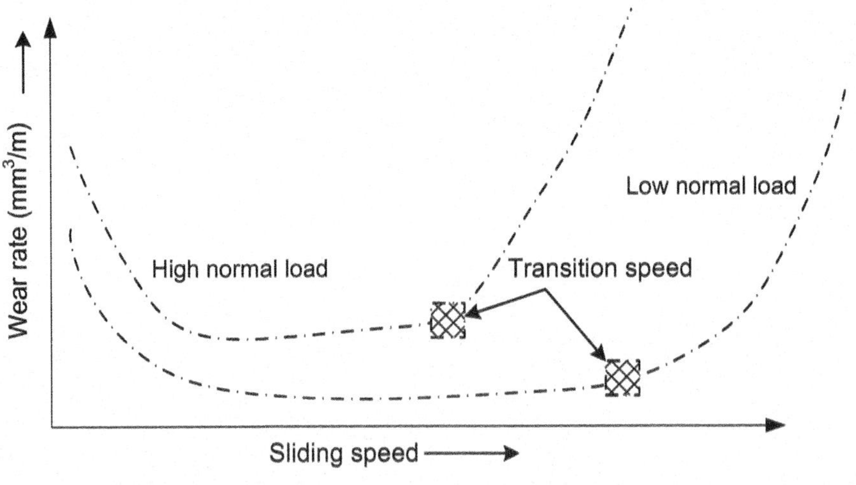

Fig. 2.6 Effect of sliding speed on wear rate at different normal loads

softened metal results in 10–100 folds or even more increase in wear rate causing severe metallic wear. The sliding speed at which transition from mild oxidative wear to severe wear takes place is called transition speed.

Relative motion between wear sample and counter surface
Relative motion between wear test specimen (pin, ball) and counter surface (disc, flat, ring) can be either unidirectional or reciprocating type. Effect on relative motion on the wear rate depends on the way it causes the presence/absence of wear debris at sliding interface (Fig. 2.7). Wear debris mostly comprises oxides and deformed and work hardened metallic particles. Debris present at the sliding interface between mating components can act as a third body to cause three-body abrasive wear. In case of unidirectional motion like pin-on-disk and ball-on-ring configurations, wear debris generally gets removed from the counter surface due to centrifugal force, and therefore, it is usually not found at the sliding interface. Chances for the presence of wear debris at the sliding interface under reciprocating relative motion condition are found more than unidirectional motion. Hence, unidirectional motion usually causes lower wear rate than reciprocating motion.

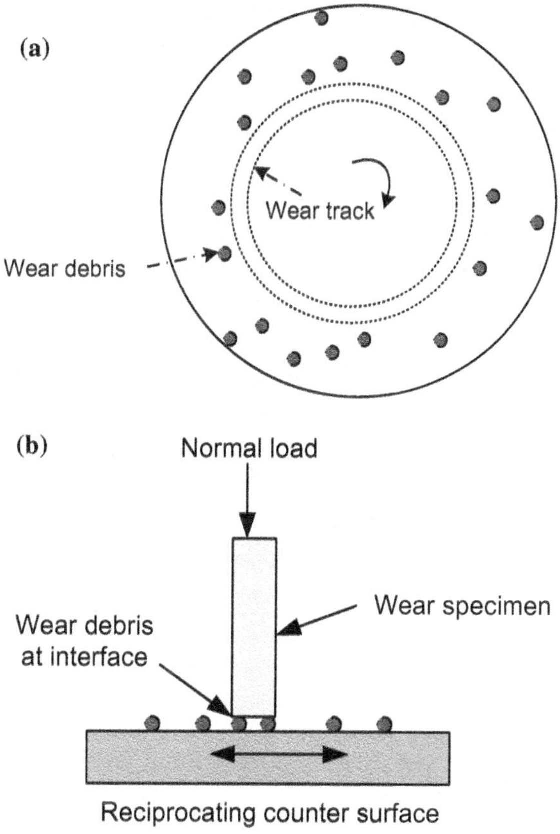

Fig. 2.7 Relative motion between wear specimen and counter surface **a** unidirectional and **b** reciprocating motion

Counter surface

The metal system against which component is made to slide is called counter surface. The composition, crystal structure, hardness, and temperature of the counter surface affect the metallic intimacy at mating interface in two ways (a) deformation of surface irregularities and (b) oxidation of interacting surfaces which in turn influence wear rate of the specimen. Similarity of composition and crystal structure of counter surface and wear specimen results in more metallic intimacy. During the sliding increased metal-to-metal contact causes higher wear rate. Therefore, counter surface (even for engineering application at the design stage) is generally selected of different compositions and crystal structures from component in consideration. Temperature of the counter surface affects both oxidation tendency and thermal softening. Slight increase in temperature of counter surface increases the oxidation of interacting surfaces which in turn lowers the wear rate due to reduced metal-to-metal contact. However, significant increase in temperature of the counter surface may lead to thermal softening and even partial melting of the softer metal. Softening of metal increases the metallic intimacy, and therefore, it experiences higher wear rate. Hardness of the counter surface also affects the metal-to-metal contact. High hardness of counter surface generally decreases the wear rate due to somewhat reduced metallic intimacy owing to limited interfacial yielding. Therefore, counter surface and mating wear component of matching hardness cause high wear rate.

Environment

The tribological components under adhesive wear conditions are required to work in different environments like ambient air, steam, gases, and liquid metal. Influence of environment on wear rate depends on how far it affects the metal-to-metal contact due to formation of oxides, nitrides, and any other metallic components at the sliding interface. The formation of hard, coherent, and stable oxides and compounds reduces the metallic intimacy between sliding components which in turn decreases the wear rate while the formation of porous, brittle, and weak layers of oxides and compounds increases the wear rate in two ways (a) formation of oxides and their removal under sliding conditions and (b) presence of impurities at sliding interface may act as third body to increase the possibility of three-body abrasive wear.

Classical model on adhesive wear

According to the simplified model developed by Archard, the volume of material (V) removed from the surface under adhesive wear conditions is directly proportional to the product of cross section (A) of the interfacial bond developed because of the plastic deformation and length of groove (D depends on sliding distance) that has formed due to the relative motion on the surface (Fig. 2.8). The cross-sectional area of the interfacial bond is primarily determined by the extent to which plastic deformation occurs in the subsurface region. The extent of deformation depends upon the applied normal load W (in N) and the flow stress ($H/3$). The flow stress is found to be approximately one-third of the hardness H (Pa) of material.

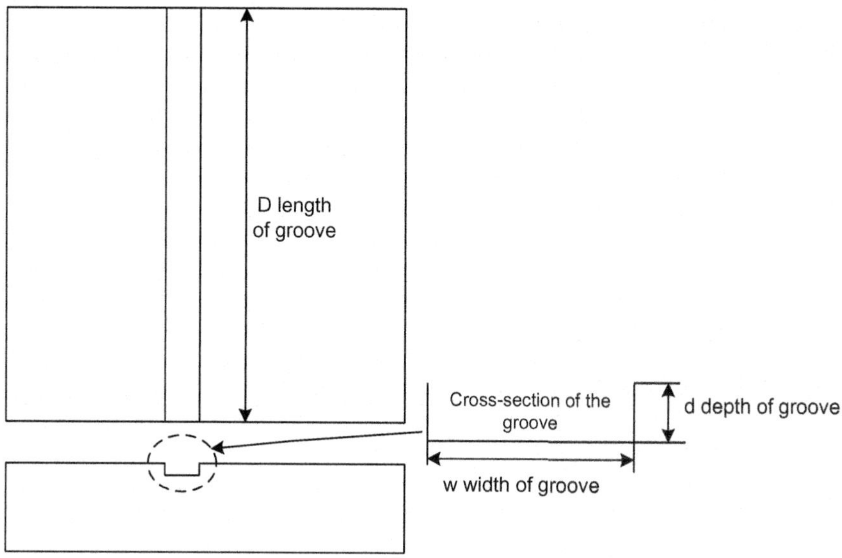

Fig. 2.8 Schematic of groove formation during adhesive wear for determining wear rate

$$\text{Wear volume } V\text{: } (k \times D \times W)/(H/3) = (3 \times k \times D \times W)/H$$

where k is the proportionality constant called wear coefficient, which is a dimensionless number, W is load (N), D is distance (m), V is wear volume (m^3) and flow stress (Hardness/3 in Pa). Wear coefficient is found to vary from 0.001 to 10^{-12} depending upon the interfacial contacts as dictated by oxidation or lubrication condition. The above equation suggests that the adhesive wear rate (*V*/*D* in g/m) of the interacting surfaces increases proportionally with the increase in normal load and decreases with the increase in hardness.

2.3.2 *Abrasive Wear*

The interaction of functional surfaces of components with hard particles like rocks, sand, dust causes one of the most significant forms of the wear called abrasive wear which in a simplified form is supposed to occur in a way that is similar to that of microcutting by the number of randomly oriented fine cutting edge of hard particles. Based on the appearance of wear surface, the abrasion of engineering components takes place through two ways, namely plowing and cutting. Technically, the mechanisms generally held responsible for material loss from the surfaces under abrasive wear conditions include plastic deformation followed by work hardening and then fatigue, grain detachment, and brittle fracture. The common abrasive wear mechanism is depicted in Fig. 2.8a–c.

Mechanisms of abrasive wear

Depending upon the hardness, toughness, and size of the hard abrasive particles as well as the above properties of abrading surfaces, it will be decided whether plowing or cutting will take place. Hard and stable abrasive particle (under given load condition) with favorably oriented sharp cutting edges removes the material by cutting action, whereas the blunted abrasive particles (without sharp cutting edges) result in plowing by displacing the material in sides and producing a wedge-shape groove. However, the formation of wedge-shape groove does not necessarily remove the material but continuous deformation work hardens it and is subsequently removed by fatigue (by coalescence of cracks) due to the repeated action of abrasive particles. In an effort to reduce wear by increasing hardness backfires after certain limit (of increased hardness) owing to the loss of toughness and increased brittleness (Fig. 2.9). Metals with limited or nil ductility or brittle metals with the action of abrasive particles develop the subsurface cracks which in turn lead to the loss of material by brittle fracture. Similarly, poor bonding of microconstituents in a material system such as in composite materials (having SiO_2, SiC, WC, Al_2O_3, etc., in aluminum matrix) leads to the debonding of hard particles from matrix and detachment of grains from materials through abrasive action. Sometimes, matrix holding the hard particles in position is abraded from all around by abrasive action, thereby leading to the loss of support to hard microconstituents which in turn results in grain detachment in the form of pullouts.

Factors affecting abrasive wear

The volume of material lost from the surface through abrasion primarily depends on the depth of penetration and relative movement of abrasive particles with respect to the indented surface. The depth of indentation is governed by many materials, service load, and abrasive particles related to parameters. In general, the increase in hardness, reduction in applied load, and size of abrasive particle result in decreasing the depth of penetration which in turn lowers the abrasive wear.

Moreover, apart from shape, the mechanical properties of abrasive particles, that is, hardness and toughness predominantly affect the abrasive wear, as these two properties determine the stability of abrasive particles in terms of size and shape under abrasive wear conditions. Hard abrasives generally get fractured and so provide sharp and fresh cutting edges for rapid abrasive wear to take place but the size gets reduced, whereas tough abrasives get blunted with the rounded-off edges and corners which in turn leads to reduced abrasive action (Fig. 2.10). So, there is a critical toughness or hardness combination resulting in maximum abrasive wear due to the balance between formation of sharp cutting edges and survival of grit size.

In case when hardness of abrasives is significantly higher than surface, abrasive wear varies linearly with the size of abrasive particles.

A ratio of the hardness of surface material and abrasive (1.2) is one of the crucial parameters affecting the abrasive wear rate (Fig. 2.11). Abrasives having the above ratio lower than 0.8 result in rapid wear. If this ratio is higher than 1.2 then it results

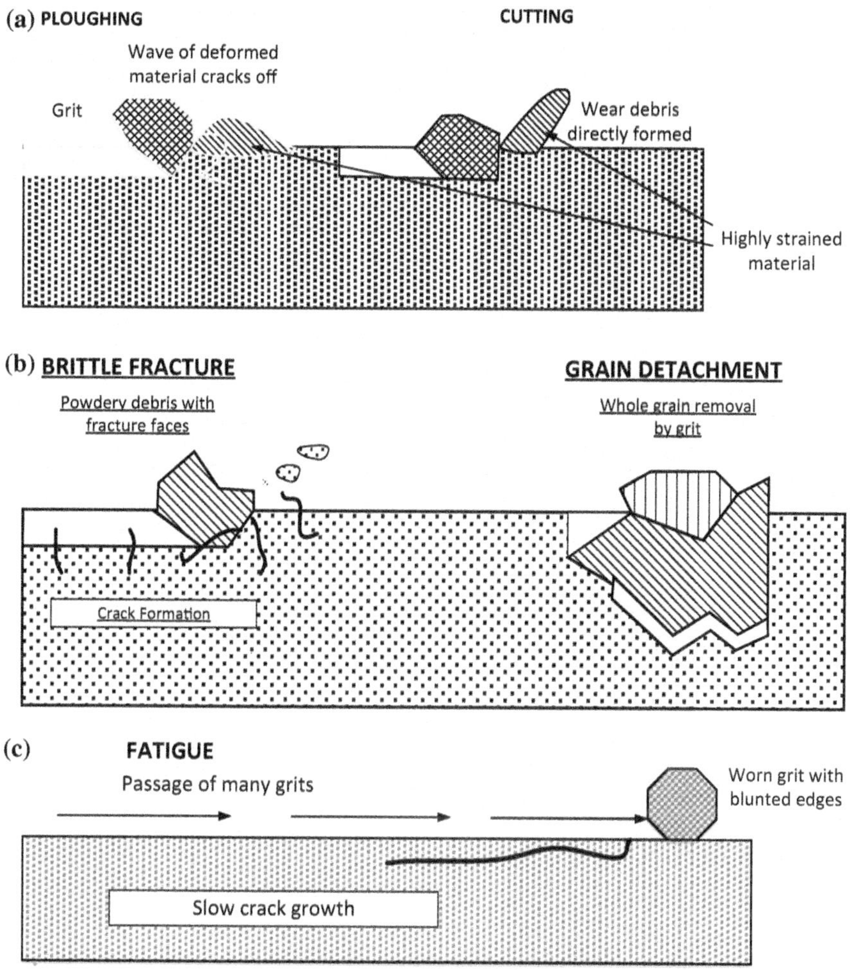

Fig. 2.9 Schematic of abrasive wear mechanisms **a** plowing and cutting, **b** brittle fracture and grain detachment, and **c** fatigue

in negligible wear. Designer always makes the effort to reduce gap in the above two harnesses for reducing abrasive wear.

Moreover, the hardness of surface during abrasion is found to be different from that of the bulk material owing to difference in the rate of straining or loading. Material under abrasive conditions is strained at much higher rate (just like metal cutting by shearing) than normal hardness tests; therefore, the hardness of severely deformed material is considered to be a more useful parameter in relation to the wear than normal hardness (Fig. 2.12).

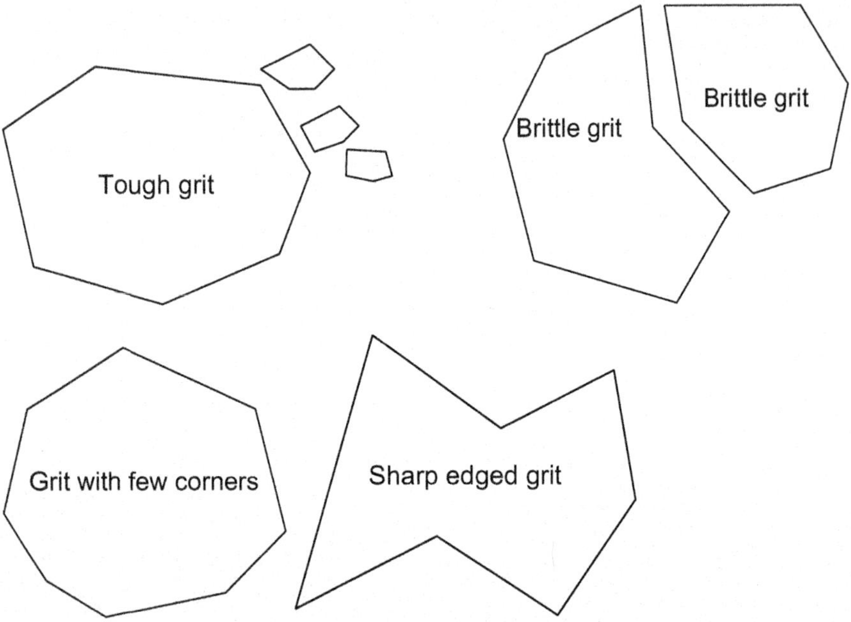

Fig. 2.10 Schematics showing the ways by which abrasive grits are affected during wear

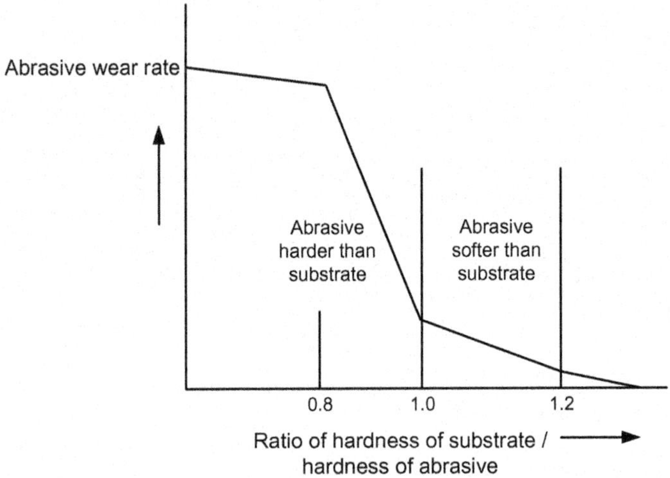

Fig. 2.11 Effect of hardness ratio of substrate and abrasive on abrasive wear rate

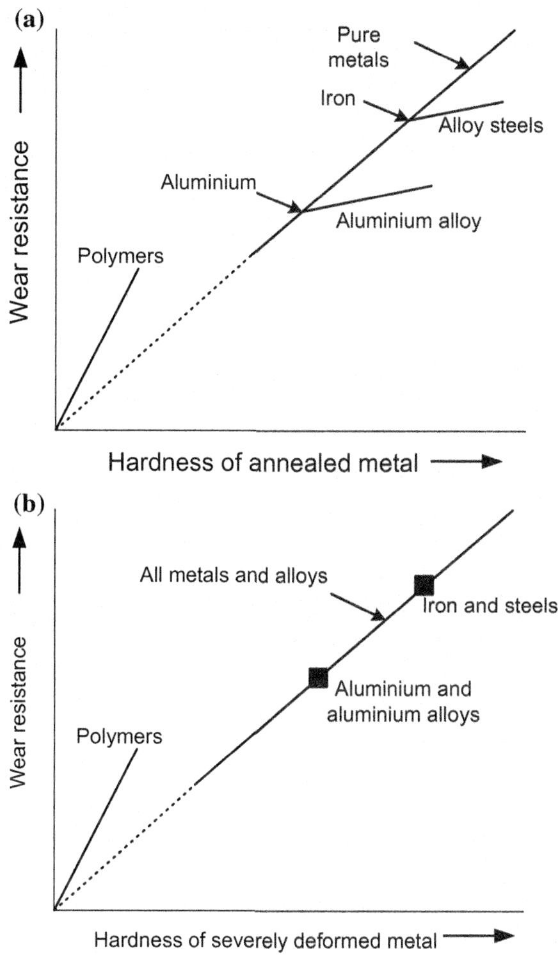

Fig. 2.12 Schematics showing the effect of hardness on abrasive wear rate of **a** under deformed material, and **b** severely deformed material

2.3.3 *Erosive Wear*

The erosion of solid surfaces can occur due to high impingement of fine particles (in solid state) or droplets moving at high speed. The mechanism of material removal owing to erosion can vary appreciably depending upon the speed, angle of impingement, size and shape of particles, phase of particle, that is, liquid or solid, and mechanical properties of the surface in consideration (Fig. 2.13). These parameters dictate the mechanisms of material loss from the surface such as shearing and cutting (like abrasion), plastic deformation, brittle fracture and fatigue, super-plastic flow or melting. According to the prevailing erosion mechanisms, the hardness, toughness, fracture toughness, and fatigue resistance of material play a predominant role in determining the life of components.

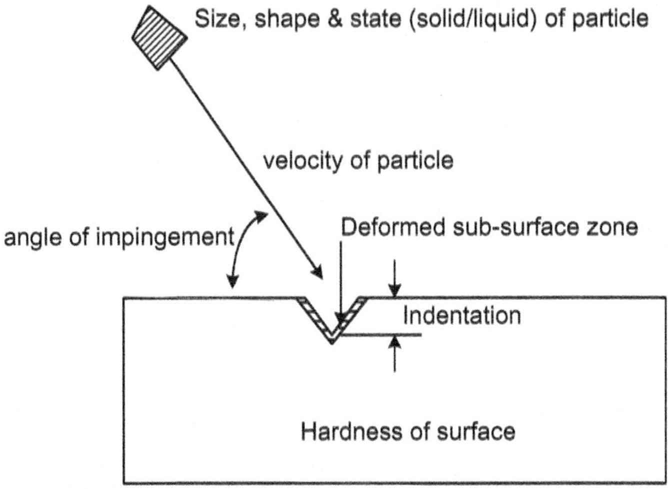

Fig. 2.13 Schematic of the erosion of surface and important variables

In erosive wear, the kinetic energy of particle at the time of impact and surface characteristics play a major role in controlling the operational mechanism of material loss. The kinetic energy of the particles during the time of impact is found to be the function of mass of particle (so size) and its velocity. In general, an increase of kinetic energy of impacting particle increases rate of erosion.

Effect of angle

Fine solid particles (about 20 μm) moving at a high speed (about 100 m/s) impinge at a very low angle at the surface remove material by abrasion like mechanisms which involve plowing, shearing, and cutting, etc. These mechanisms can also be termed as a ductile mode of erosive wear. Shearing and cutting will occur only if some indentation is caused by the impinged particles because of the kinetic energy during the time of impact. A reduction in angle of impact reduces this energy, and so the related indentation is also decreased which in turn lowers the erosion (Fig. 2.14). Particles impinging at a high angle produce high kinetic energy during the time of impact (maximum at 90°), which results in the loss of material from the surfaces of tough metals through the combination of plastic deformation and fatigue, whereas in case of high strength and low ductility metals loss of material takes place through brittle fracture. For brittle metals, the maximum impact energy of particles at 90° causes the highest erosion rate and the minimum erosion is observed for those impacting at a low angle. For ductile metals, maximum erosive wear is observed at about 30° of the angle of impingement.

Particle size

Atomic-scale size particles (ions and atoms) moving at very high speed cause surface damage by removing the atoms from the lattice, while in the case of large-size particles (more than 50 μm), melting and super-plastic flow predominantly determine

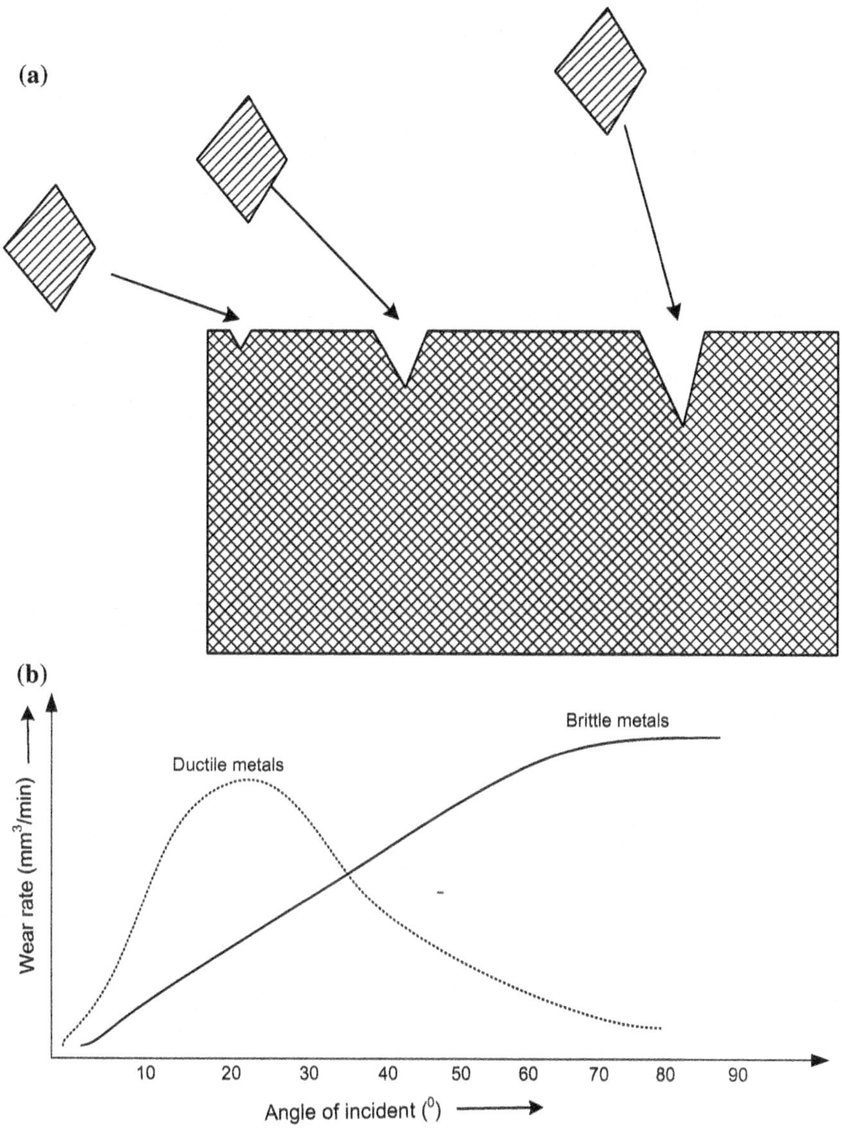

Fig. 2.14 Schematic showing the effect of angle incidence on **a** indentation and **b** wear rate during erosion

the erosive wear rate. At a moderate speed, the small-sized particles (about 10 μm) having low kinetic energy at the time of impact result in the deformation and work hardening of thin subsurface layer, which in turn makes it crack-sensitive and prone to removal, while the large-size particles make larger volume of subsurface layer deform up to a greater depth that is sensitive to cracking and erosion. Therefore, fine particles encourage the ductile mode (owing to fact that the limited volume of subsurface layer is subjected to lesser plastic deformation), while the large particles promote the brittle mode of erosion. High kinetic energy of large particles moving at a very high speed generates excessive heat at the time of impact which results in the melting of a thin surface layer. Part of this molten layer is lost (with particles or splashing sidewise) from the surface, and the rest of it re-solidifies on the surface.

Substrate materials

Material properties—hardness, toughness, and ductility—predominately affect the operational erosion mechanism which in turn affects the erosion rate (Fig. 2.15). Hard, high strength, and low ductility of materials resist indentation which in turn affect the removal of material by erosion. The particles impinging at low angle to the surface cause erosion through plowing and cutting mechanism, where the first indentation is a prerequisite so any phenomenon or modification (such as alloying, overlaying, work hardening) increasing the surface hardness would reduce the wear by erosion. Erosion by solid particle impinging at high angle is predominantly caused by brittle mode (through crack nucleation and growth mechanism). Any mechanism or factor enhancing the fracture toughness or toughness of surface layer (in case of high ductility and low yield strength metals) would increase resistance of the brittle fracture and high angle erosion, as the increasing hardness beyond a

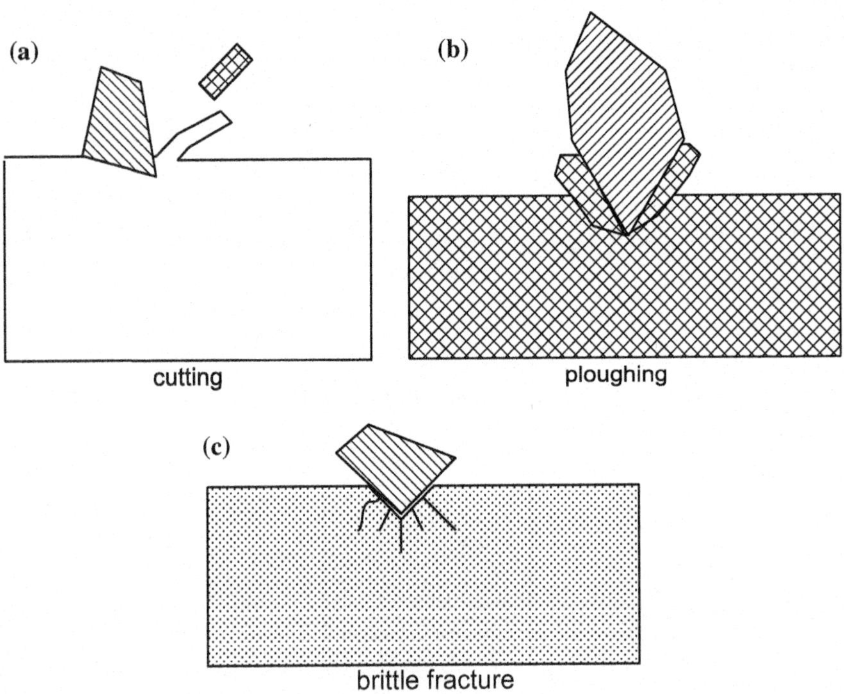

Fig. 2.15 Schematic erosive wear mechanisms: **a** cutting, **b** plowing, and **c** brittle fracture

certain level would only deteriorate the erosion resistance. Therefore, increase in toughness with reasonable level of hardness lowers erosion by particles impacting at high angle.

Particle shape
Particle with sharp edges and corners helps in erosion by cutting and shearing, while spherical particle lowers the erosion wear. Erosion by particles in a liquid state is found to be very sensitive to particle velocity, and erosion is found to be a proportional to the fifth power of velocity.

Melting wear
The development of abnormal and extreme sliding conditions such as excessive normal load, sliding speed, and failure of lubrication leads to the generation of frictional heat at sliding interface. The rise in temperature results in the melting of any of the mating parts which cause melting of thin surface layers (up to few micrometers depth). Increase in the surface temperature affects the wear and friction behavior in two ways: (a) thermal softening and (b) interaction of metallic surface with the surrounding atmospheric gases. The influence of both on the wear and friction characteristics with rise in surface temperature is found to be opposite. Thermal softening with limits (before melting) facilitates the plastic deformation of surface asperities, so as to increase the metal-to-metal contact area which in turn tends to increase both friction coefficient and wear. However, the interaction of surface layer with surrounding gases (oxygen, nitrogen, etc.) at high temperature results in the formation of oxides and nitrides in accordance with the affinity of surface metal with gases present. The formation of oxides layer at the surface generally performs two functions: (a) reducing the direct metal-to-metal contact between the interacting surfaces, thereby reducing the friction coefficient and wear rate by adhesion and (b) developing the hard protective mechanical mixed surface layers which is generally harder than subsurface layer (Fig. 2.16).

The mild increase in interfacial temperature under designed or service low load, sliding speed, and lubrication condition results in mild oxidative wear conditions, whereas the abrupt rise in interfacial temperature (beyond certain critical level) due to accidental increase in load, speed, and lubrication failure leads to localized melting of near-surface layer up to the depth of few micrometers. The presence of thin films of molten layer at the sliding interface decreases the forces required for maintaining the relative motion (owing to poor/negligible interface bonding), which in turn decreases the friction coefficient. However, the peeling of molten metal layer from the interface under the influence of normal load leads to rapid increase in the rate of material loss from the interacting surfaces (especially one having low melting point and resistance to thermal softening). The loss of material from the functional surfaces takes place in the form of big metallic wear particles. This increase in wear rate under melting wear conditions can be 10–100 folds or even greater than that of mild oxidative wear conditions.

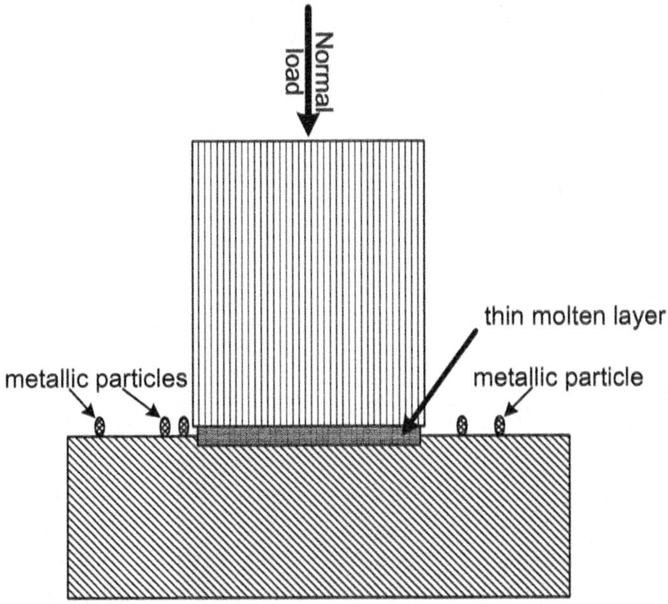

Fig. 2.16 Schematic of the surface erosion by melting wear

2.3.4 *Corrosion Wear*

The loss of material from the functional surface owing to interaction (in form of chemical reaction) between surface and fluid present around it is called corrosion wear. The nature of reaction product on the surface in the form of oxide, nitrides, etc., determines the rate of material loss (wear) from the surface of metals. A coherent, adherent, and non-porous reaction product in the form of oxides (e.g., aluminum oxide, chromium oxide) results in significantly low rate of wear because these oxides do not allow the access of oxygen in the air to react with the metal below the oxide layer, thus preventing or reducing further oxidation or corrosion. However, the formation of porous, loosely held, and non-coherent reaction products (e.g., iron oxide in the form of rust) formed by the interaction of metal with corrosive fluid results in high rate of material loss from the functional surfaces through corrosion, as these products of reaction allow the access of corrosive media to the metal surface apart from getting dislodged easily from the surface (Fig. 2.17). Any activity such as the interaction of affected metal surface (having corrosion reaction product) with other components, bodies, abrasives, cavitation, fluids, etc., that can remove the reaction products formed on the functional surface will expose the fresh metal surface to corrosive media for further chemical reaction or attack on the surface, which in turn will eventually increase the material loss from the surface by corrosion. For example, the adhesive or abrasive wear in the corrosive environment will have synergic effect of material loss from the functional surfaces by

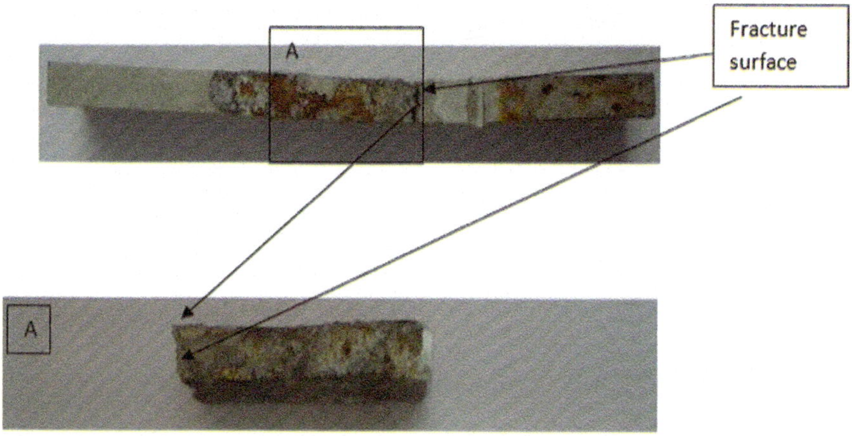

Fig. 2.17 Photograph of corroded surface of FSW joints of Al alloys in 5% NaCl solution

causing material loss through both adhesion or abrasion and corrosion. Wear rate under such condition is found to be many times greater than the combined wear by both the mechanisms, individually. Metallic components working in the corrosion environments like marine, mining, petrochemical environments are found sensitive to such kind of synergic effect of wear and corrosion.

2.3.5 *Diffusive Wear*

High normal load and sliding speed conditions lead to direct and intimate metal-to-metal contact between the interacting surfaces (like in seizure conditions) which results in the atomic movement (diffusion) of alloying elements from one region or body or surface to another depending upon the presence of driving force in the form of concentration gradient of alloying elements or energy level. However, there are few requisites for the occurrence of any influential diffusion of alloying element from one body or region to another, and these requisites are (a) perfect metallic intimacy, that is, the absence of impurities or oxides layers between the interacting surfaces, (b) high temperature which is enough to have reasonably good diffusion coefficient, and (c) contact duration between interacting surfaces must be long enough for diffusion. The relative movement between mating components under abnormal or normal load and sliding speed can develop such conditions at the interface leading to the diffusion of alloying elements. The diffusion of alloying elements can create areas or zones in the interacting surfaces having unfavorable enrichment or depletion of alloying elements which in turn can deteriorate their mechanical, corrosive, and tribological properties. Wear of metal cutting tool during machining is a typical example of diffusion wear.

It is well known that during metal cutting, peak temperature is generated near the cutting edge (at some distance away from the cutting edge) on the rake face of the cutting tool against which chips sliding under high pressure create a direct metal-to-metal contact (Fig. 2.18a). The presence of high temperature and metal-to-metal contact between the rake face of tool and underside surface of chip facilitates diffusion, especially in the absence of proper cooling. Diffusion wear significantly reduces the life of tools that are made of common materials like high speed steel, tungsten carbide-cobalt, polycrystalline diamond primarily during the machining of carbon steel under the unfavorable machining conditions (cutting speed

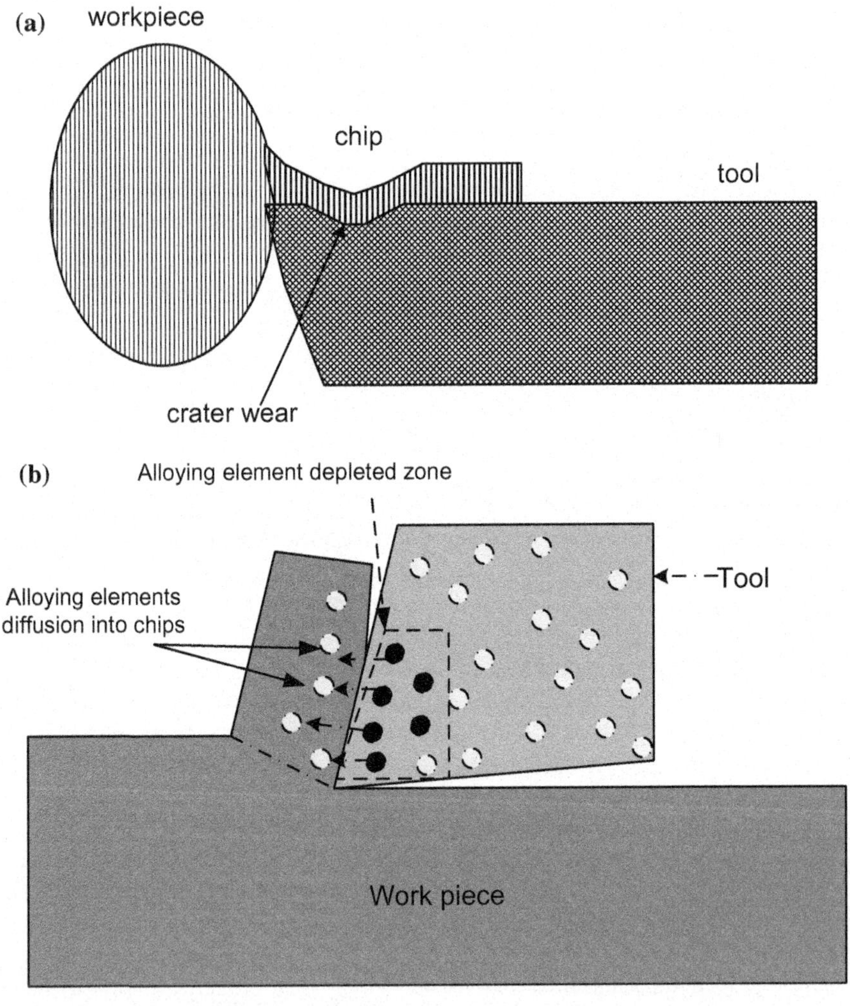

Fig. 2.18 Schematic of diffusive wear of **a** tool during metal cutting and **b** diffusion of elements from cutting tool

beyond recommended or specified level and use of excessively high feed rate and depth of cut) lead to high heat (interface temperature) generation and pressure at the cutting edge to facilitate diffusion of alloying elements from the tool surface and subsurface layer to sliding metal chips. The loss of alloying elements (W, Cr, C, etc.) from the rake/flank face of high-speed steel tool and the carbon from the surface of polycrystalline diamond tool during machining of steel reduces high temperature stability and resistance to thermal softening (Fig. 2.18b). Softening of surface and near-surface layers of cutting tool causes the loss of material from the face of cutting tool by adhesion, as cut metal chips slid over the face of the tool under pressure.

2.4 Techniques to Evaluate Damage of Wear Surfaces

Uncontrolled and progressive loss of material from the surfaces of mating components having sliding or rolling contacts influences the properties of surface and subsurface layers with respect to the roughness, composition, microstructure, mechanical, and corrosion properties. However, the extent of influence of the above characteristics depends on wear mechanism which is responsible for surface damage. For example, the adhesive and surface fatigue wear results in changes in the form of weight loss (dimensional changes), roughening of surface, work hardening of surface and subsurface layer, severe straining and microstructural modification of near-surface layers, formation of mechanical mixed layer (having oxides of element of both sliding components). The quantification related to the wear surface damage is mainly carried out to establish (a) the length of time necessary for the occurrence of changes in the sliding components in terms of dimensional variation and surface roughness that can lead to failure, (b) the steps need to be taken at the design stage of component for reducing the rate of surface degradation so as to enhance the useful service time, and (c) the development of strategy for reclamation or refurbishing of worn-out component.

The characterization of each type of features of surface damage due to wear needs separate and specialized approach. The most common methods of studying the wear consist of the examination of sliding material before and after the test; any difference in material is attributed to wear. The detection of wear generally uses one or the other techniques of weighing, mechanical gauging, and examination of surface and subsurface features and wear debris.

2.4.1 Material Loss

Amount of the material lost from the wear surface is quantified using the three common approaches that are (a) weight loss, (b) height or depth measurement, and (c) optical method. The last method is employed when the first two methods cannot be applied in a situation.

2.4.2 Weighing

This is the simplest way of detecting the wear in which specimen is weighed before and after running test by using sensitive weighing balance (accurate > 0.1 mg), and weight loss is calculated to get wear rate.

2.4.3 Dimensional Measurement

In this method, the wear is measured through the decrease in dimensions using mechanical (dial gauge) or electrical system based on the linear variable displacement transducer (LVDT) principle. This has resolution limit of about 10^{-5} m.

2.4.4 Optical Method

There are a number of methods for measuring wear using the optical technique. One way is to make small microhardness indentation on a surface and to study how its size is reduced during sliding. The horizontal limit of resolution of this method is about 10^{-5} m.

2.4.5 Hardness of Surface and Subsurface Layers

Vickers microhardness test is generally carried out using low load (say 100 g) for obtaining hardness profile from the surface to subsurface regions. It indicates the depth up to which the deformation has taken place during wear.

2.4.6 Chemical Composition and Phase Analysis

Spectroscopy and X-ray diffraction analysis are commonly performed to establish the properties with respect to the chemical composition and phases present on the surface generated (e.g., mechanical mixed layer) during service. The presence of elements and their compounds in the form of oxides, nitrides, and chlorides from both interacting components confirms the formation of mechanical mixed layer. The presence of surface compound further confirms the existence of particular kind of environments during service. For example, the formation of Fe_2O_3 and Fe_3O_4 indicates the occurrence of corrosion of iron in the presence of oxygen in moist environment. This information can also be used in the failure analysis for establishing the root causes of failure which in turn can be used to avoid failure from that particular mechanism. The formation of hard, strong, and coherent protective mechanical mixed surface layer is considered to be good from the point of view of wear and friction resistance.

2.4.7 Surface Roughness

The roughness of surface subjected to wear is very dynamic characteristics and provides enough information for drawing inference regarding the mode of wear and operating wear mechanisms. However, surfaces subjected to different wear conditions like adhesion, abrasion, erosion, cavitation, corrosion exhibit significant variation with respect to the surface morphology. Typical surface morphologies and their features produced under different wear conditions are shown in Fig. 2.19a–d. Adhesive wear refers to the presence of oxides, and the scoring marks in the direction of sliding besides the presence of cracks observed at high magnification on the wear surface suggest the occurrence of mild oxidation wear, whereas bright shining surface with heavy plastic deformation of grooves and ridges along with the development of large metallic debris suggests the occurrence of severe metallic wear. Severe metallic wear becomes 10–100 times greater than mild oxidative wear. Abrasive wear shows deep and wide abrasive and scratch marks on the surface which suggest the occurrence of high rate of material removal. Erosion refers to cavitation, slurry, and solid particle erosion of the subjected surface that commonly reveals the presence of surface, pits, cracks, and craters. The depth of craters determines the wear rate by these processes.

Questions for self-assessment

1. How can performance of an engineering component be related with service life?
2. Describe common causes of deterioration in performance of material.
3. What are the common types of wear experienced by metals?
4. Explain the mechanisms of adhesive wear.
5. Describe factors affecting adhesive wear.
6. How do service conditions affect the adhesive wear?
7. Explain classical law of adhesive wear.
8. What is abrasive wear? Describe mechanisms of abrasive wear.
9. Explain the factors affecting abrasive wear of metals.
10. How do material properties affect the erosive wear?
11. Explain mechanism of erosive wear.
12. Describe the factors affecting erosive wear of metals.
13. Explain melting wear with the help of schematic diagram.
14. What is diffusive wear? Explain the mechanism of diffusive wear observed in cutting tool.

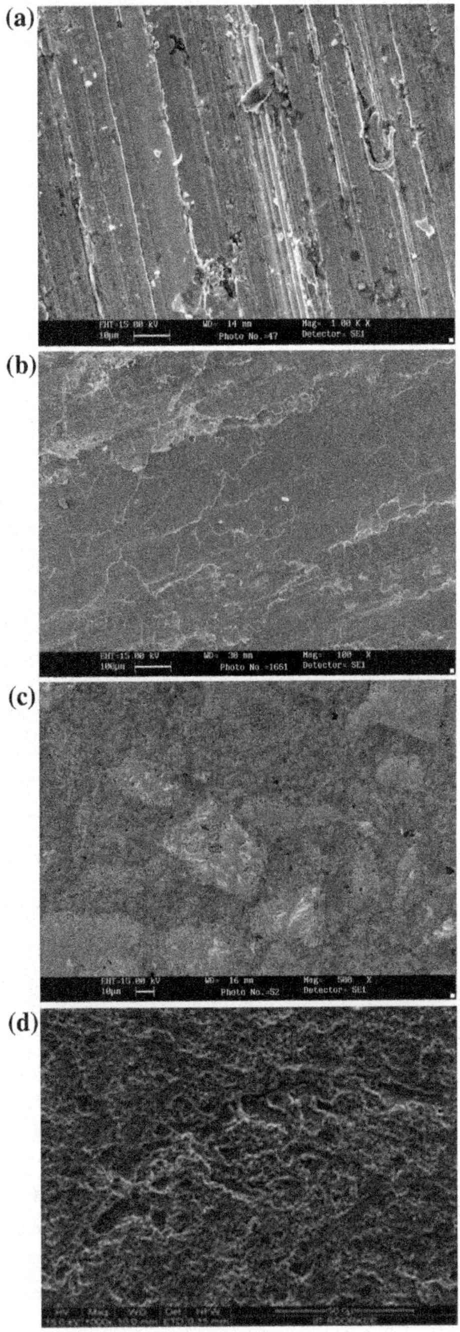

Fig. 2.19 SEM micrographs showing typical surface features of samples subjected to **a** abrasive wear, **b** adhesive wear, **c** slurry erosion, and **d** cavitation erosion

15. Describe method used to measure the wear.
16. What is significance of surface and subsurface studies of worn-out samples?
17. How can effect of wear on surface and subsurface of worn-out sample be studied?

Chapter 3
Materials for Controlling the Wear

3.1 Materials Properties and Wear

To control the wear, functional surfaces must possess certain mechanical (hardness, toughness, yields strength, ductility, fracture toughness, resistance to thermal softening), chemical (affinity or resistance to surrounding medium atmospheric gases, lubricant), metallurgical (microstructure, stacking fault energy), and dimensional properties (presence of notches, roughness, and waviness). The importance of the required common properties of the surface is highlighted below. Moreover, the relative importance of these properties varies with service conditions which contribute toward the loss of materials from the surfaces.

3.1.1 Hardness

Hardness indicates the resistance to indentation which is indirectly determined by the flow stress or yield strength of a metal. Low hardness of metals facilitates (a) The formation of deep and wide indentation and plastic flow during erosion, (b) The formation of deep and wide scratch and easy shearing of metal through cutting mechanism during abrasion, (c) Plastic deformation of surface asperities and near subsurface layer increases the metallic intimacy during adhesion related to wears such as adhesive wear, fretting wear, surface fatigue wear. On the basis of the above views, it can be stated that moderate increase in hardness is always considered to be good from the wear resistance standpoint. However, increased hardness leads to decrease of other useful and desirable material properties such as ductility, toughness, notch, and fracture toughness that are needed to reduce the material loss by various wear mechanism. Conversely, the decline in other properties owing to excessive increase in hardness can severely lead to reduction of the materials' wear resistance against surface fatigue, abrasion, erosion, and cavitation.

D. K. Dwivedi, *Surface Engineering*,
https://doi.org/10.1007/978-81-322-3779-2_3

Further, hardness of a functional surface must be designed keeping in mind the hardness of mating component surface, abrasive or erosion medium, etc., as the ratio of hardness of the considered material and the counter surface or particle significantly determines wear rate.

3.1.2 *Ductility*

The ductility of a metal system facilitates the deformation under external stresses without cracking. The deformation of peaks and valleys present at the surface affects the wear in many ways because the increased metallic intimacy and mechanical interlocking increase the adhesive wear and friction coefficient.

3.1.3 *Toughness*

The toughness of a metal system is the combination of strength and ductility which indicates the ability to take up the shock or impact loads without fracture. A high toughness metal needs high stress for nucleation and growth of cracks to cause fracture. It plays an important role in determining the abrasion, adhesion, and erosion of surface metal having fatigue wear resistance, especially those coupled with impact loading. Metal system for solid particle erosion and cavitation resistance as well as surface fatigue wear conditions must have an optimum combination of hardness and toughness.

3.1.4 *Stacking Fault Energy*

The stacking fault energy associated with a metal system (owing to fault in arrangement of atoms in lattice) directly affects the work hardening tendency, which in turn determines the hardness and yield strength of the surface and subsurface regions at the time of imparting service under external load condition. The increase in strength and hardness takes place when surface or subsurface regions are subjected to plastic deformation, and the same is attributed for enhancing resistance to the movement of dislocations by cross-slip. In general, low stacking fault energy metals (gold, copper, stainless steel) work hardens more rapidly than high stacking fault energy metal systems (nickel and aluminum). The stacking fault energy predominantly determines the wear behavior of surfaces that are subjected to wear and are coupled with the surface fatigue, abrasion under impact, cavitation, solid particle erosion, etc. Therefore, Mn steel and Mn–nickel steel, cobalt base alloy having low stacking faulty energy offer good resistance to wear under impact conditions.

3.1.5 Fatigue Resistance

The fatigue resistance of metal indicates the number of load cycles it will take to complete all the three stages: (a) crack nucleation, (b) stable crack growth, and (c) fracture of component or surface layer. The most of wear mechanisms such as surface fatigue wear, erosion, and cavitations involve fluctuation or repeated loading on the surfaces.

3.1.6 Fracture Toughness

The fracture toughness of a metal system shows the resistance to crack growth to cause fracture. Fracture toughness is important especially under those wear conditions when the removal of material from the surface takes place through crack nucleation and growth mechanisms like surface fatigue wear, erosion, and cavitation. High fracture toughness decreases the crack growth rate under external stresses and delays the removal of materials by delamination and brittle fracture.

3.1.7 Affinity to Atmospheric Gases

Metals forming adherent and hard protective layer, develop a very stable and coherent mechanical mixed layer which tends to reduce the friction coefficient and wear by avoiding direct metal-to-metal contact. Moreover, the formation of brittle and non-coherent oxides that are removed rapidly during sliding between mating components results in high wear rate. The nature and characteristics of surface oxide layer play an important role in determining the friction and wear behavior under the mild sliding wear conditions rather than severe sliding conditions.

3.1.8 Thermal Stability of Surface Layers

During the sliding between the mating components, frictional heat is invariably generated which increases the temperature of the sliding interfaces. The rise in temperature (10–15 °C) is very limited under mild wear conditions; therefore, changes in structure, strength, and hardness owing to frictional heat generation are expected to be negligible. Hence, thermal stability (in respect of mechanical properties and structure) of the wear surface layers under mild wear conditions does not play a significant role, whereas thermal stability of a wear surface under severe sliding conditions (during rise in temperature due to localization of the frictional surfaces) becomes more significant. Poor thermal stability of metals at high temperature can lead to thermal softening, which in turn can reduce their ability to take normal load during sliding, and this will eventually lead to severe wear condition. The resistance to galling, high temperature erosion, and diffusive wear is especially determined by thermal stability of the metal under consideration.

3.2 Materials Properties Required for Better Wear Resistance

The mechanisms contributing toward the material loss from the surface depend on the nature of contact, materials in consideration, and service conditions. The loss of materials from surface by each mechanism is controlled by certain properties of material. The properties of metals important for controlling the different kind of specific wear are described in the following section.

3.2.1 Adhesive Wear

Adhesive wear resistance of the material is governed by the characteristics of surface layers with respect to ductility, hardness, yield strength, microstructure, and nature of oxides formed during sliding due to affinity with atmospheric gases. As far as possible, it is advisable to avoid mating or sliding between similar kinds of metal systems because of high tendency in forming the metallic bond between them at the sliding interface, which in turn results in high friction coefficient and wear rate. Increase in hardness results in high wear resistance of surfaces under dry sliding conditions, mainly when load is largely static, and there is no possibility of cyclic loading. The presence of complex microstructure and non-metallic microconstituents in the metallic matrix reduces the tendency of electron transfers at the sliding interfaces, which in turn reduces friction and wear of metal from the surface because of the decreased incidences of metallic bond formation. However, the presence of very hard microconstituents in the matrix sometimes increases the wear by roughening the counter surface due to abrasion by hard particles.

Galling Wear
Under severe load and high speed sliding conditions, the generation of high friction heat due to almost complete metallic contact between the mating surfaces that leads to excessive thermal softening of surface and near surface layers, which in turn deteriorates the hardness and ability of carrying the applied load during service. These conditions result in high friction coefficient and wear rate. The resistance to thermal softening and stability of the surface layers under severe sliding conditions primarily dictate the galling resistance of a metal-counter surface combination, which in turn is determined by ductility, stacking fault energy, and crystal structure of metal systems. Metal systems having low ductility, low stacking fault energy,

and crystal structure (like BCC, HCP) with fewer slip systems generally resist cross-slip, as all these factors help in increasing the resistance to plastic deformation. Low stacking fault energy metals decrease the cross-slip tendency and hence increase the resistance of plastic deformation. The metals in ascending order of stacking fault energy are: gold, copper, nickel, and aluminum.

3.2.2 *Abrasion*

The resistance to the abrasion of a metal system is influenced by properties, namely modulus of elasticity, yield strength, ductility, hardness, fracture toughness, composition, and microstructure in terms of size, shape, and interparticle spacing between hard microconstituents; these properties can affect the loss of material by plowing, cutting, and brittle fracture. In general, low stress abrasion wear is mainly controlled through hardness, whereas high abrasion wear resistance is significantly determined by work hardening behavior of the materials in question. Further, materials having fine and uniformly distributed hard particles with close spacing in the matrix yield better abrasion wear resistance than that of coarse and widely spaced particles.

3.2.3 *Erosion*

Loss of materials from the functional surfaces by erosion can be caused by the impingement of solid particles carried by some fluid (liquid or gas) and bursting of gas bubbles formed in liquid handling pumps or pipelines under low pressure conditions. The shock loading of surfaces either by (a) the impact of solid particles at high velocity or (b) the pressure waves generated by repeated bursting of bubbles primarily causes plastic deformation followed by work hardening, nucleation of cracking, and eventually removal of material through the coalescence of cracks. In general, the nucleation of cracks is preceded by plastic deformation (may be very small); therefore, it is important to have moderate hardness besides good toughness for absorbing the impact of particles or pressure waves. The hardness of surfaces must be closely controlled else it would deteriorate the toughness and increase the erosion wear by brittle fracture. All the factors namely work hardening, fatigue and corrosion resistance, and fine microstructure, etc.—discourage crack nucleation and retard their growth, will lead to enhancement of the resistance to erosion. Further, the reinforcement of hard particles in matrix of soft material does not help much in enhancing the erosion resistance because soft matrix is selectively attacked by both under solid particle erosion and cavitation.

3.2.4 Surface Fatigue Wear

Material characteristics such as hardness, fracture toughness, work hardening tendency, and microstructure affect crack nucleation, and growth which in turn influence the surface fatigue wear. In general, an increase in surface hardness reduces the plastic deformation, which decreases the tendency of crack nucleation at the particle–matrix interfaces both at the surface and subsurface regions. However, high hardness does not help much in improving the surface fatigue wear resistance because of the fact that increase in hardness negates the advantage of delayed crack nucleation owing to increased crack growth rate, as the tip of crack growing in hard metals remains sharp which causes high stress concentration that facilitates the crack growth at faster rate, whereas in the case of low hardness (high ductility and low yield strength), crack tip gets blunted, which reduces the crack growth rate and so more time is required for the removal of material from surface in the form of debris through delaminating mechanism. Similarly, metals of high fracture toughness and good work hardening tendency improve the surface fatigue wear. In general, fine grain structure having spherical morphology offers higher resistance to surface fatigue wear (due to the requirement of high stresses to nucleate cracks and their growth) than coarse grains and needle shape microconstituents. Thus, a good combination of surface hardness and facture toughness becomes a requisite for good surface fatigue wear resistance.

3.2.5 Cavitation

The material loss by cavitation mechanism takes place through bursting of bubbles formed during the flow of fluid on the functional surfaces of many engineering components, such as impeller of pumps, valves, hydro-turbine blades. The pressure produced by bursting of bubble can be as high as 1500 MPa, which is enough to cause localized plastic deformation that is followed by work hardening and eventually crack nucleation and formation of crater and cavitation, after the coalescence of cracks. All the factors that can delay nucleation and growth of cracks under the influence of shock waves of bursting bubbles will enhance the resistance of the cavitation. For cavitation resistance, metals with high fatigue resistance, fracture toughness, good combination of hardness yield strength and ductility, low stacking fault energy, and low surface roughness are preferred. High hardness can delay the nucleation stage of crack during the incubation period, but subsequently high crack growth rate reduces the life of component subjected to cavitation. Therefore, a combination of hardness and toughness certainly helps in achieving high cavitation resistance.

3.2.6 *Fretting Wear*

The oxidation of metal at the surface of substrate and then removal of oxides from the functional surfaces under fretting conditions (fluctuating loading due to oscillatory motion between the matting surfaces) is believed to be controlled by (a) the formation of adherent and coherent oxides of metals and (b) the development of good resistance to cracking because of better mechanical properties with respect to fatigue, fracture toughness, yield strength, and ductility. This in general offers better fretting wear resistance than those forming loose and non-coherent oxides and is crack-sensitive. These requirements are based on the fact that material removed under fretting wear is caused by cracking and oxidation, and thereafter, fracture develops minute metallic pieces or oxides.

3.3 Selection of Materials for Surface Engineering

There is no single metal system that can satisfy the property requirement under all conditions of wear; therefore, it becomes mandatory to analyze the various service conditions related to the parameters dictating the loss of material from the surfaces for establishing the mechanical, chemical, and metallurgical characteristics which should be achieved at the tribological surfaces for the increased wear resistance and longer life of component in consideration. Commercially, the different following metal systems are commonly used for engineering the surfaces.

3.3.1 *Iron Base Alloy*

Iron base alloys offer good hardness, tensile strength, toughness, moderate resistance to corrosion, both high and low temperature applications especially when weight is not a major factor (owing to high density of these metal systems)

- Carbon and alloy steel;
- Stainless steel;
- Ferritic, pearlitic, and martensitic irons;
- High chromium and carbon irons.

3.3.2 *Cobalt Base Alloys*

Cobalt base alloys provide a good combination of hardness and toughness in relation to work hardening capability and resistance to thermal softening.

3.3.3 *Nickel Base Alloys*

Nickel base alloys offer high toughness and resistance to high and low temperature conditions.

3.3.4 *Copper Base Alloys*

Copper base alloys are used for better corrosion resistance.

3.4 Structure and Wear of Material

Microstructure of metals and composites can be understood better by looking into their phase structure and grain structure. The phase structure indicates (a) the type of phases, (b) their relative amount, and (c) the distribution of phase in matrix. The grain structure indicates (a) the size of grain, (b) the shape of grain, and (c) their distribution in matrix. These six features of microstructure of any metal system determine their response to the external loading and behaviors under different conditions like wear, corrosion. Therefore, it would be of utmost importance to be aware of the microstructural aspects that are related to metals in consideration for understanding their response to tribological conditions and planning the ways of making modification for achieving target properties such as improved resistance to wear, corrosion, fatigue.

In general, most of the alloys and composites are designed to have a combination of well-distributed hard constituents (like cementite, martensite, carbide and boride of chromium, eutectic mixtures, tungsten, vanadium, $CuAl_2$, Mg_2Si, silicon) in the matrix of tough and comparatively soft metal (ferrite, aluminum, titanium, cobalt, nickel, etc.) for the enhancement of mechanical and tribological performance of bulk materials, overlays, and coatings. A metal system (of given composition) having coarse, needle-shaped, and widely spaced hard constituents offers lower wear resistance (under adhesion, abrasion, erosion, cavitation, etc.) than one with fine, spherical, well-distributed, and closely spaced hard constituents. The fine, closely spaced, and spherical-shaped hard particles being embedded in the matrix of soft and tough metal, there is the requirement of higher stress to initiate and develop voids or cracks at hard particle–matrix interface than those having coarse, widely spaced, needle-shaped particles. Easy nucleation and subsequent growth of cracks eventually produce the fracture and wear particle in the form of debris. The influence of above morphological characteristics of soft and toughness matrix on wear and mechanical properties is found inferior to that of the hard constituents. Moreover, the refinement of hard as well as soft and tough microconstituents in a metal or composite definitely results in the significant enhancement of mechanical and tribological properties. In some of the cases like manganese steel and Hadfield steel, the deformation assisted phase transformation from austenite to martensite during service like impact abrasion, cavitation, erosion, results in the continuous improvement in mechanical and wear properties. Therefore, the utilization of suitable approaches for controlling the morphology of hard microconstituents in a favorable manner results better mechanical and tribological performance.

3.4.1 Ferrous Metals

Ferrous metals include alloy systems like wrought alloys, steel, and cast irons, and these are of great commercial importance. These are mostly composed of common phases like ferrite, cementite, austenite, martensite, graphite, and fine and intimate mixture of above in different forms like pearlite (lamellar ferrite and cementite), ledeburite (austenite and cementite), bainite (fine ferrite and cementite). Each phase or phase mixtures has its own mechanical properties. The above-mentioned phases can be found in different shapes such as lamellae, needle, spherical (compact or irregular), flake, cuboid, polyhedral, star, and Chinese script. Each shape has different impact on the mechanical properties like fatigue, fracture toughness, tensile strength, and ductility, which in turn governs the wear behavior under different service conditions.

3.4.2 Carbon Steel

Carbon steels are primarily produced with controlled amount of carbon that is considered as an alloy element, whereas other residual elements like Mn (<1%) and Si (<0.6%) and S and P (<0.05%) are controlled. The typical microstructure of carbon steel shows ferrite and pearlite (Fig. 3.1a). Carbon (wt%), in plain carbon steels, directly affects the amount of ferrite and pearlite in the hypoeutectoid range (up to 0.8% of C) as well as the amount of pearlite and cementite in the hypereutectoid range (0.8–2.0%). Increase in pearlite and cementite in carbon steels in general increases the hardness but at the cost of toughness and ductility (Fig. 3.1b). Fine pearlite, cementite (in the form of fine discrete particles), and martensite in quenched and tempered (Q and T) conditions result in better wear resistance than coarse and lamellar structures.

Therefore, low carbon steel is used for application of surfaces, where ductility and toughness are more important than hardness, whereas high carbon steel is used for the application that requires high hardness and abrasive wear resistance.

More et al. (*Wear* 28 (1974) 59–68) reported that increase in fraction of pearlite (10%) in ferritic steel increases abrasive wear resistance, and the same was

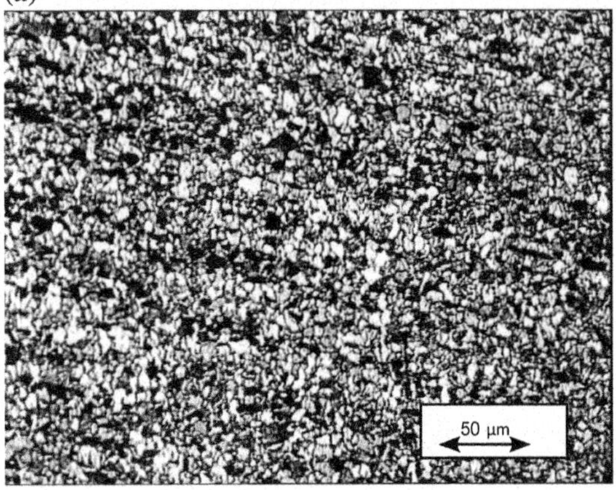

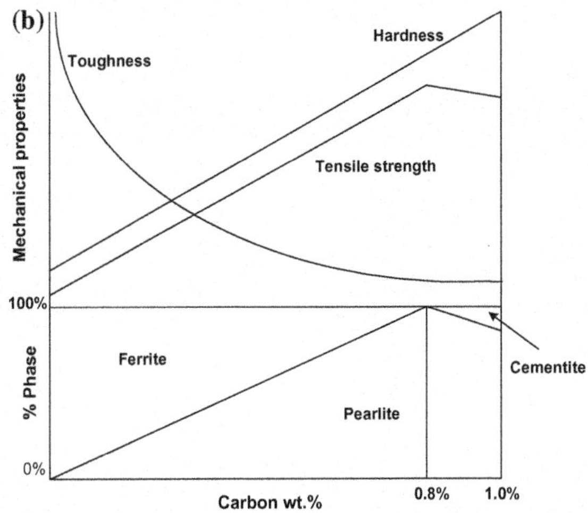

Fig. 3.1 **a** Optical micrograph of mild steel. **b** Schematic showing the effect of carbon (wt%) on phase structure and mechanical properties of plain carbon steel

attributed to two factors: (a) Increasing fraction of hard phases restricts the movement of abrasives during wear and (b) lowering the plastic flow of soft ferrite matrix. Therefore, increase in relative amount of well-dispersed hard pearlite in the comparatively soft ferrite matrix increases the abrasive wear resistance irrespective of abrasive grit size.

3.4.3 *Alloy Steel*

Alloy steel generally contains small amount (<1%) of alloying elements such as Cr, V, W, Al, Mo, and Ni for achieving the specific properties with respect to the hardness, toughness, high temperature resistance etc. (Fig. 3.2). These elements

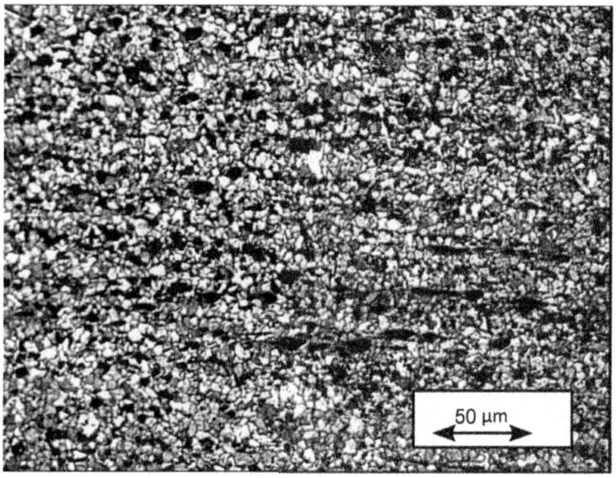

Fig. 3.2 Optical micrograph of low alloy steel

increase the resistance to thermal softening (yield strength) and corrosion. Refinement of grain structure by the addition of aluminum and vanadium in steel and the formation of hard and stable carbide of W, Cr, and Mo increases the hardness and wear resistance.

3.4.4 Stainless Steel

Stainless steels are mainly of four types, namely, austenitic stainless steel, ferritic stainless steel, precipitation hardening stainless steel and martensitic stainless steel. Among these stainless steels, probably, the austenitic stainless steel is the most commonly used steel for corrosion resistant applications, whereas the martensitic stainless steel is preferred for wear-resistant applications in corrosive environments such as hydro-turbine components, cutlery, and surgical tools. The presence of hard martensite in martensitic stainless steel offers desired resistance to deformation, and high percentage of chromium (>12%) in these steels results in requisite corrosion resistance. However, the formation of undesirable chromium carbide during slow cooling from 650 to 850 °C causes sensitization which in turn leads to localized loss of the corrosion resistance especially along the grain boundary areas, where the formation of chromium carbide develops Cr-deficient areas. Chromium-deficient areas act as anode for preferential attack under favorable corrosive conditions. This in turn leads to many undesirable effects such as weld decay, stress corrosion cracking, and intergranular corrosion.

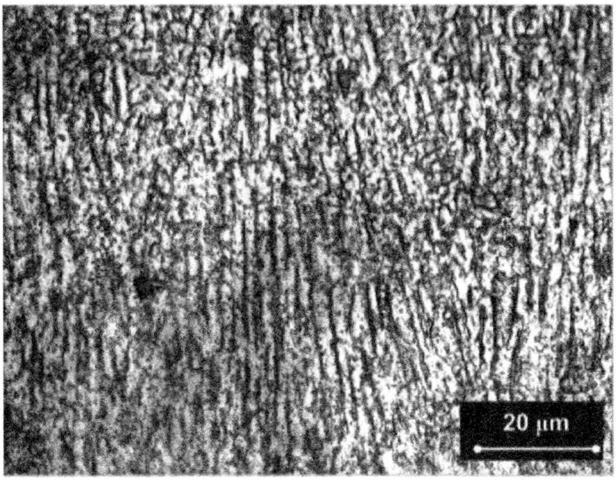

Fig. 3.3 Optical micrograph of Hadfield steel

3.4.5 Hadfield Steel

Hadfield steel is a special steel having high manganese (12–14%) and nickel designed to have austenite in the matrix which during service when subjected to localized deformation under external load hardens readily (as high as 63 HRC) owing to stress-assisted transformation of austenite into martensite (Fig. 3.3). This feature of Hadfield steel is exploited in the applications such as cavitation condition and abrasion coupled with the impact for having improved tribological life of component in consideration.

3.4.6 Gray Cast Iron

Gray cast iron depending upon the composition and cooling conditions experienced during solidification can be composed of graphite flake in the matrix of ferrite, pearlite, and bainite. Thus, matrix significantly determines the hardness of cast iron because graphite is comparatively soft and acts as solid lubricant under adhesive wear conditions. Moreover, graphite reduces the transfer of vibration through metallic matrix because its flakes break the continuity of metallic matrix.

3.4.7 White Iron

White iron is a type of cast iron that is formed when the iron experiences high cooling conditions during solidification. These are mainly composed of iron carbide in the matrix of pearlite or martensite. The presence of these hard constituents results in extremely high hardness which makes white iron suitable for low stress abrasive wear conditions. However, poor toughness and ductility result in unsuitable wear conditions that are coupled with impact. Hard iron carbide reduces the indentation and depth of abrasive marks under low stress abrasive wear condition which in turn improves wear resistance against abrasion.

3.4.8 Chromium Irons

Chromium iron is a type of cast iron that frequently contains high percentage of chromium and carbon. Chromium iron is basically a Fe–Cr–C system containing various types of carbides, namely M_3C, M_7C_3, Cr_7C_3, and Fe_3C. Each carbide has different composition and possesses different hardness. M_3C predominantly contains Fe_3C with some amount of Cr. It is primary carbide having <15% Cr. M_7C_3 is basically Cr_7C_3 with some amount of iron. It is also a primary carbide with >15% Cr. Hardness of Cr_7C_3 (2100 DPH) is more than Fe_3C (1300 DPH). During abrasion, soft matrix is selectively abraded which leads to easy removal of hard carbide particle as these are left without matrix support. However, increase in concentration of hard carbide particles in the soft matrix increases the load-carrying capacity and abrasion resistance (Fig. 3.4) (Chatterjee and Pal 2003).

In Fe–C–Cr–Nb–Mo system, formation of niobium carbide particles refines the chromium carbide particle through heterogeneous nucleation which in turn improves the abrasive wear resistance due to reduced spalling tendency of fine carbide particles. Further, higher hardness of niobium carbides than chromium carbide in comparative soft matrix of austenitic also results in reduction of low stress abrasive wear. The improvement in low stress abrasive wear resistance is not found to be appreciable beyond certain fraction of hard carbide particles owing increased fragmentation tendency of brittle carbides. A good combination of hardness and tough matrix is considered to offer good low stress abrasion resistance.

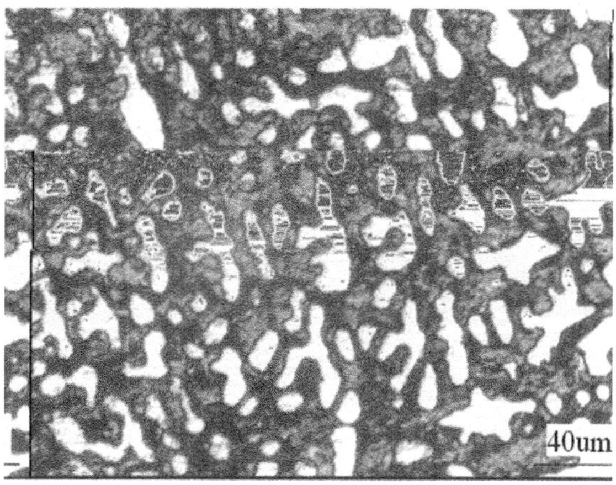

Fig. 3.4 Optical micrograph of high Cr and high C iron

Effect of Common Alloying Elements on the Performance of Fe based Hard Facings

The type of the alloying element and its percentage affect the wear resistance of iron-based hardfacing. Influence of various alloying elements in iron on the mechanical and microstructural characteristics has been summarized below.

Effect of Boron

Boron is mostly used in steels in order to increase hardenability; that is, to increase the hardening depth during quenching. Boron-treated steels usually contain a boron content in the range of 0.0005–0.003%. This small amount of boron appears to be the optimum range for enhancing the hardenability of other alloys. Boron is designated as an alloy "intensifier." Boron addition is very effective with low carbon alloy steels, and effect of boron reduces with increase of the carbon content. If carbon content is above 0.60%, the boron addition is not recommended. Boron adversely affects impact resistance of steels at low temperatures as it increases ductility to brittle transition temperature (DBTT).

Boron is one of the alloying elements, which can be used in place of carbon to form chromium and iron borides. Borides possess high hardness and have high potential of good abrasive wear resistance. Boron can be a substitute for the costlier alloying elements like niobium or titanium. The boron addition in Fe-based austenitic steel enhances the grain boundary strength and refines the austenite grain size. The addition of boron provides low melting point eutectic phases that promote coherency between carbides, borides, and matrix material during fusion. The boron addition increases adhesive wear resistance and hardness. The addition of 2% boron exhibited higher hardness due to the formation of primary coarse hard phases like iron–boron (Fe–B) type boride. Increase in boron content promotes the growth of primary Cr_7C_3 and increases the volume fraction of hard phases, which in turn enhances hardness and the wear resistance. The morphology of hard phases affects the wear behavior significantly.

Effect of Carbon

Carbon is an essential element to the formation of cementite (as well as other carbides) and to the formation of pearlite, spheroidite, bainite, and martensite in Fe–C systems. The hardness of steel (and hardenability) increases with addition of carbon up to 0.65%. The wear resistance of iron based system can be improved with carbon addition up to about 1.5%. Further, increase of C decreases toughness and promotes brittleness significantly. The carbon percentage affects carbide formation and the matrix phase of the deposits. The carbon content is reported to be most significant variable in determining the low stress abrasion resistance. To obtain better wear resistance, high hardness of the hardfacing is needed; therefore, high content of C and carbide-forming elements is required in hardfacing fluxes for enhanced hardness and wear performance of material. The formation of carbides, their size, and volume fraction depends upon the amount of carbon content being added. Large fraction of M_7C_3 carbides are formed at higher carbon concentration. The increased volume fraction of carbides and the formation of M_7C_3 primary carbides at higher carbon results in improved wear resistance. The volume fraction of austenite increases with the increase of carbon, manganese, and nickel which in turn results in a significant decrease in hardness. The use of graphite in iron-based hardfacings has also been reported in some of the studies. Graphite is an allotrope of carbon and known to be the most stable form of carbon. The amount of graphite should be controlled within 10–12% because further increase in quantity of graphite, continuous cementite and retained austenite is formed (Fig. 3.5). The increase of cementite decreases the cracking resistance of alloys. The matrix having lesser

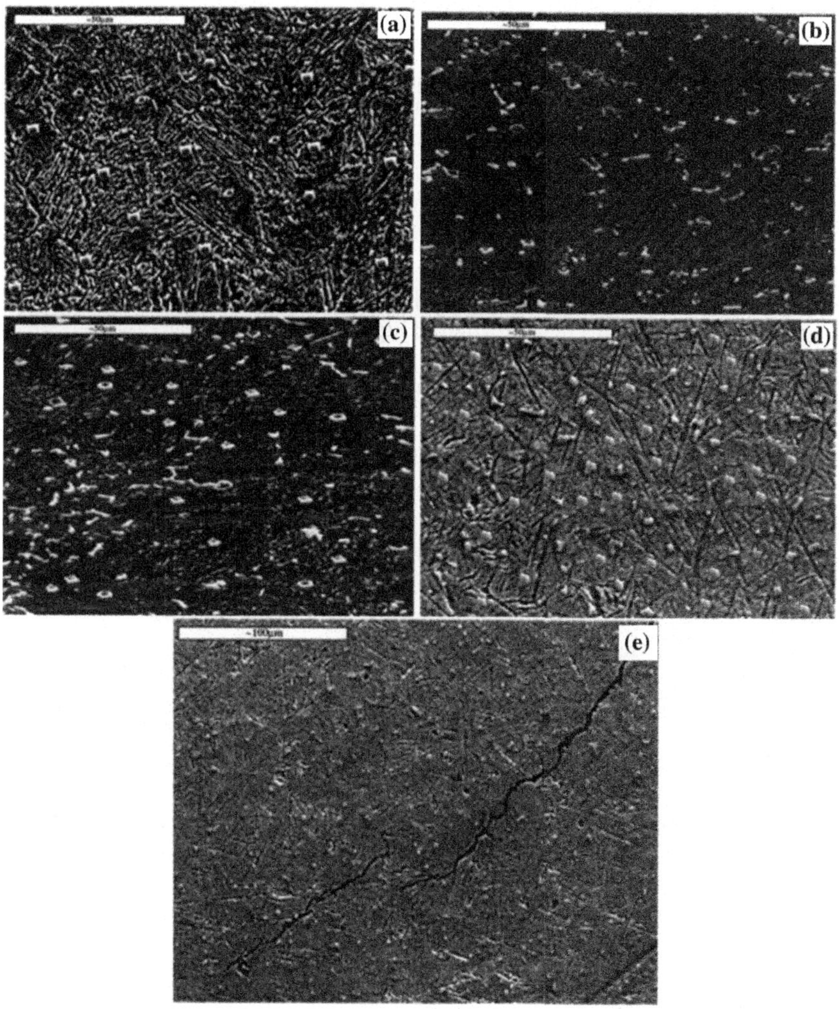

Fig. 3.5 Influence of the graphite on the microstructure of hardfaced weld deposits: **a** 6% graphite; **b** 8% graphite; **c** 10% graphite; **d** 12% graphite; **e** graphite over 12%, cracks appeared (Wang et al. 2008)

than 5 wt% graphite causes deficiency of the carbides because amount of graphite is not found enough to react with titanium, vanadium, or molybdenum to form carbides.

Effect of Chromium

Chromium addition improves the hardenability of steel. It can also improve the toughness and the wear resistance of steels along with many other interesting effects. Addition of 5% Cr or more with Mn reduces the critical cooling rate and makes the steel air-hardenable. Cr increases resistance to staining and corrosion. The steels containing 12% or more chromium are known as stainless steels. Steel with chromium increases the critical temperatures for heat treatment. Iron–chromium–carbon hardfacings with high Cr content have been extensively used commercially as hardfacing materials in many industrial applications (which includes mining, cement plants, thermal power plants, and iron and steel industries). Among these, white cast irons with a high concentration of Cr and C are extensively used because of the low cost and superior wear resistance. High carbon, high chromium irons show 20–25 times more abrasion resistance than mild steel under two bodies and three-body abrasive wear conditions against comparatively softer abrasives than carbide particles. The high hardness and the excellent wear resistance of high Cr-C-Fe based alloys are attributed to the formation of the chromium carbides. Chromium does not only act as a ferrite stabilizer but also as a carbide former, especially when C concentration is more than 0.3%. Chromium-carbide-based composites are extensively used as tribological materials in high-temperature applications requiring high wear and corrosion oxidation resistance. Hardfacing alloys with 10–30 wt% chromium and 2.0–3.5 wt% carbon content possess the high volume fraction of M_7C_3 carbides with high hardness. $(Cr, Fe)_7C_3$ carbides are also found in iron–chromium–carbon alloys with higher contents of carbon (2–5 wt %) and chromium (18–30 wt%). The internal stresses are relatively high in straight chromium–carbon type system because acicular shape chromium carbide promotes stress concentration. Therefore, change in the shape of carbide while retaining its hardness, abrasion resistance, and density enhances impact resistance. This has been achieved by adding Nb which in turn results in rounded Nb/Cr carbide shape. Nb containing high chromium irons with W and Mo give good wear resistance at 600 °C.

Effect of Manganese

Manganese is generally found in all commercial grades of steel. Mn improves hot working characteristics by negating the adverse effect of sulfur. Mn forms solid solution in both alpha and gamma iron. It lowers the eutectoid percentage of carbon (e.g., 2.0% manganese results in eutectoid at 0.67% C) and eutectoid temperature. In alloy steels, manganese decreases the critical cooling rate and increases the hardenability. Manganese content between 11.0 and 14.0% and carbon content of 1.00–1.40% makes the steel fully austenitic which offers good resistance to the abrasive wear coupled with impact stresses as Mn in iron increases work hardening-ability under high stress or severe impact loading condition. However, Mn in iron does not affect low stress abrasion resistance appreciably. Austenitic manganese steels contain min. 12% Mn with C, and other elements like V and austenite stabilizers like Ni tend to increase the toughness. Typical applications of such steels include blades for crusher, mining equipment, and rollers. Manganese addition in iron increases impact wear resistance primarily due to strain-induced martensitic transformation behavior of iron-based alloys especially at elevated temperature. To enhance high-temperature wear resistance min. 10 wt.% manganese must be added. The addition of 0.3 wt% boron with 10 wt.% manganese in iron-based hardfacing improves resistance to cavitation erosive wear.

Effect of Molybdenum

Molybdenum in steel can lead to the formation of (a) solid solution with the ferrite and (b) complex carbide structure. In alloy steels, molybdenum (0.10–0.60%) increases the upper critical point and lowers the critical cooling rate to enhance the hardenability. Molybdenum steels need a higher tempering temperature than plain carbon steel for similar level of hardness/softness. This resistance to tempering suggests ability of such steels to retain strength at high temperature. Alloy steels with 0.15–0.30% Mo show a minimum susceptibility to temper embrittlement. Molybdenum can also increase the strength via precipitation hardening at 400 °C by forming MC or M_2C carbides. The hardness and the volume fraction of carbides of the hardfacing alloys increase with the increase of molybdenum content which in turn increases the resistance to wear (Fig. 3.6).

Effect of Niobium

Niobium strengthens the steels for high-temperature applications. Nb is one of the most effective carbonitride-forming elements which in turn enhances the hardness and wear resistance. Addition of niobium increases the strength of ferritic stainless steel at elevated temperature by forming solid solution. However, prolonged exposure to niobium alloyed steel decreases the high-temperature strength due to formation of the coarse Fe_3Nb_3C. With the niobium addition, carbides in the hypereutectic high carbon high chromium irons are refined.

Effect of Silicon

Addition of silicon raises the critical temperature of steels. Silicon forms a solid solution with the ferrite. Silicon addition up to 2.5% increases the strength of the ferrite without loss of ductility. Further, larger quantity of silicon reduces machinability and increases susceptibility to decarburization. Silicon is used as deoxidizer in steel making process. Silicon increases hardenability and strength of the low alloy steels. The resistance to corrosion, oxidation, and erosion generally increases with the addition of Si in metals including steels. Silicon is known for its very low solubility in carbides (M_7C_3). Addition of silicon alters the morphology of primary carbides. Increase in silicon concentration changes morphology of the primary carbides from elongated shapes to a more equiaxed shapes which in turn improves the toughness, resistance to abrasive wear of alloy steels.

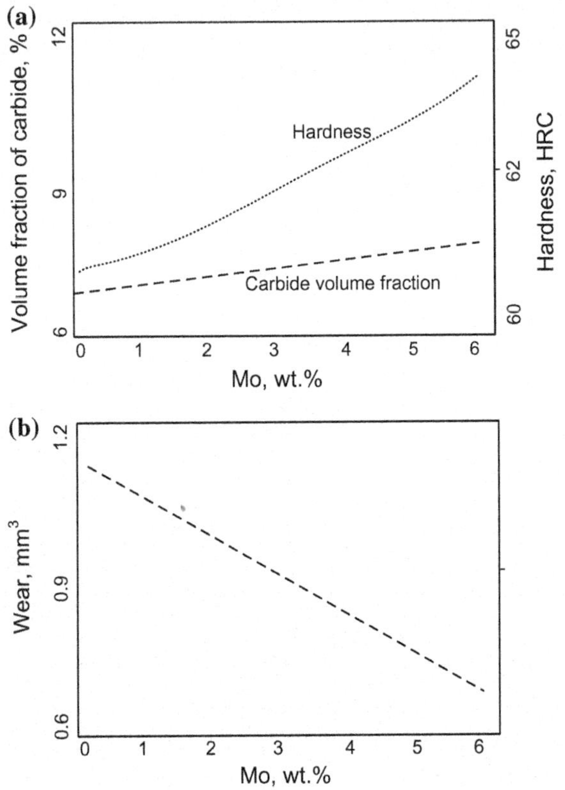

Fig. 3.6 Influence of molybdenum on **a** hardness and carbide volume fraction and **b** wear hardfacing (Wang et al. 2008)

Effect of Titanium

Ti is primarily used as a deoxidizer and an effective grain growth inhibitor in steel. Ti is a strong carbide-forming element. Effect of titanium on hardenability is marginal. Titanium is used in steel for achieving high rupture strength and secondary hardening effects. In low carbon steels, titanium strengthens ferrite by forming solid solution. Titanium residuals (of 0.02%) can adversely affect impact resistance by increasing the ductile to brittle transition temperature and increasing susceptibility to temper brittleness in steels. Further, titanium forms carbides with other alloying elements such as vanadium, chromium, niobium, and molybdenum. Titanium carbide increases resistance to plastic deformation and abrasion. Additionally, the higher hardness (3200 HV) and thermal stability of the ceramic particles of titanium carbide than that of chromium carbide make them suitable to modify iron-based alloys. Increase of titanium above a certain limit can cause problems in the slag detachability and fluidity in the weld deposits.

Effect of Vanadium

Vanadium is one of the strong carbide-forming elements. V imparts strength and toughness to steel by forming solid solution with ferrite in limited quantity. The complex carbides formed by vanadium are quite stable. Vanadium steels show a much finer structure than those without vanadium. The grain growth tendency is minimized for temperatures in the heat treating range which in turn permits higher hardening and normalizing temperatures. V raises the critical temperature and reduces the carbon content of the eutectoid point. Vanadium increases hardenability, secondary hardening effect during tempering, and hardness at elevated temperature. Vanadium addition in iron-based hardfacing results a good combination of hardness and toughness. Vanadium carbides have high hardness of 2800 HV and are formed in ternary iron–vanadium–carbon alloy. The vanadium addition refines the carbides into cast iron and cast steels. The studies related with vanadium additions in iron–chromium–carbon-based hardfacing exhibited that vanadium refines microstructure and improves wear resistance.

3.4.9 Non-ferrous Metals

Hypereutectic Al–Si Alloys and Al–Si–SiC Composites

The wear and friction behavior of Al–Si alloys are influenced by phase and grain structure of microconstituents present in these alloys. Thermal stability of phases in Al-Si alloy also affect the wear and friction behaviour appreciably. As per the silicon content, Al–Si alloys may contain alpha-aluminum solid solution, eutectic, and primary silicon (Fig. 3.7). Silicon content in these alloys has significant effect on wear resistance. However, the influence of silicon on wear properties is determined by sliding conditions such as applied load, relative velocity with counter surface, temperature. Size of silicon particles and their morphology significantly affect the sliding wear resistance of these alloys. Coarse silicon particle of high aspect ratio provides easy site for nucleation and growth of cracks at particle–matrix interface. Fine silicon particles well distributed in soft matrix of aluminum result in much better wear resistance than alloy having coarse and segregated silicon particles. The structure of alloy having hard particles in soft and ductile matrix is modified near the sliding surface due to thermal and mechanical stress developed during sliding. Hard particles become granulated in subsurface region due to localized deformation. Moreover, bonding between the hard particle and soft matrix also affects the wear behavior of such alloys. Increased bonding between hard particles with matrix increases the wear resistance irrespective of sliding condition. Spheroidization of eutectic silicon needles and improved bonding with matrix after heat treatment of Al–Si alloys improve the wear resistance.

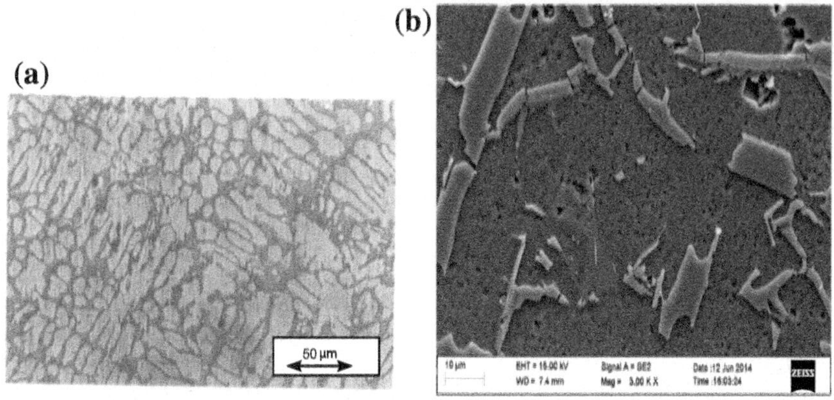

Fig. 3.7 Optical micrograph of **a** hypoeutectic and **b** hypereutectic Al–Si alloys

3.4.10 Cobalt Base Alloys and Composites

Cobalt base alloys for surface engineering application are designed to have good hardness and work hardening capability under the influence of external impact condition like abrasion with impact, cavitation. Stellite, a one typical cobalt base alloy, is extensively used for enhancing the cavitation through the application of weld surfacing, thermal spraying, and laser cladding. A combination of alpha cobalt, eutectic carbide mixture, and primary carbide of Cr and W in these alloys helps in realizing the desired hardness, abrasion, and cavitation resistance (Fig. 3.8). A laser clad of Stellite shows fine dendrites and cells of Cr–Co solid solution in the matrix of eutectic carbide mixtures, thereby suggesting that it is a hypoeutectic composition of alloy. A close look of cladding at higher magnification exhibits the presence of planar interface, followed by cellular, columnar dendritic, and equiaxed dendritic structure.

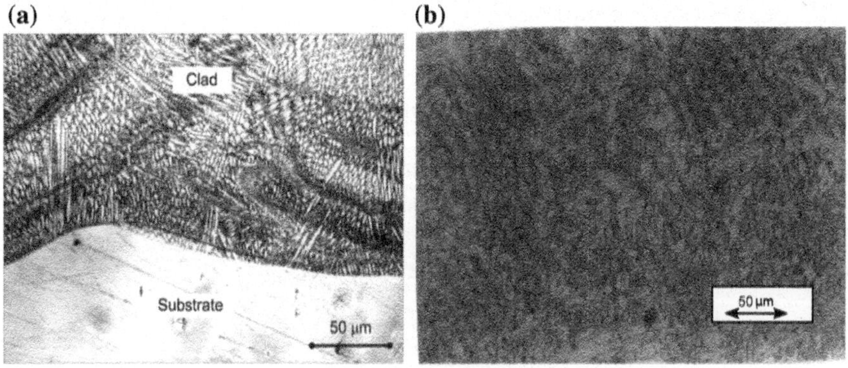

Fig. 3.8 Optical micrograph of Stellite at **a** coating–substrate interface and **b** coating

3.4.11 Nickel Base Alloys

Three commonly used nickel base alloys are (a) Ni–Cr–B–Si, (b) Ni–WC, and (c) Ni–Cr–B alloys (Fig. 3.9). Ni–Cr–B–S alloy is used for high-temperature oxidation-resistant applications. This alloy is mainly composed of nickel cell and dendrite with eutectic mixture of nickel and chromium boride which provides desired hardness and resistance to oxidation. Other two alloys namely Ni–WC and Ni–Cr–B—exhibit abrasion, erosive, and adhesive wear-resistant conditions especially at elevated temperature. The presence of hard primary tungsten carbide particles in Ni–WC imparts desired hardness and resistance to abrasion. Chromium boride particles in colmonoy (a typical Ni based alloy) result in desired hardness and wear and cavitation resistance.

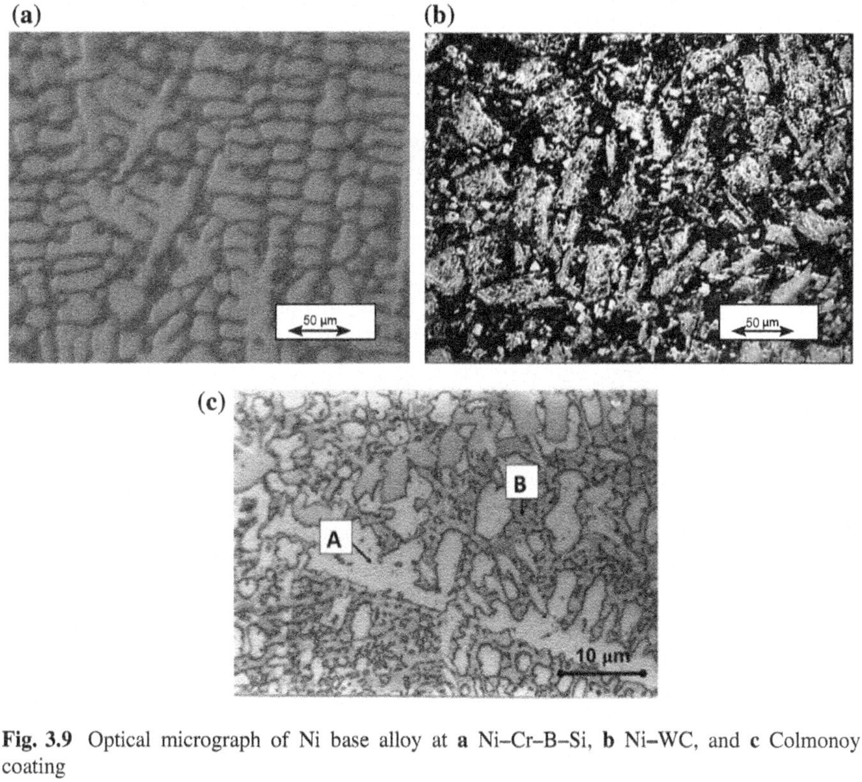

Fig. 3.9 Optical micrograph of Ni base alloy at **a** Ni–Cr–B–Si, **b** Ni–WC, and **c** Colmonoy coating

As a general guideline, we can use the following for selection of materials.

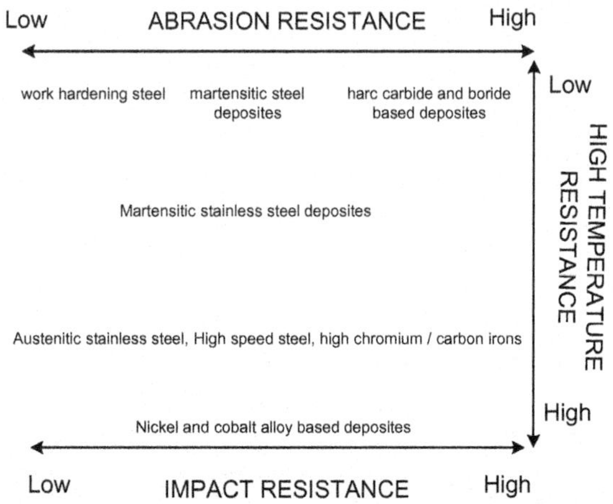

Many coating systems have been developed over the years for a variety of engineering application ranging from the enhancement of the resistance to corrosion, abrasion, and cavitation, besides increase in the thermal insulation and conductivity. These coating materials can be crystalline or amorphous and are used in form of powders, wires, electrodes, etc. These are applied over the functional surfaces using a wide range of processes such as welding, thermal spraying, laser cladding. Depending upon the application and purpose, suitable wear-resistant coating, for surface modification is selected (thermal barrier coating and functionally graded coating).

Wear-Resistant Coating Systems

The wear-resistant coating materials are mostly designed to have hard microconstituents in comparatively tough matrix. These hard microconstituents may be developed either in situ conditions through suitable mechanical, thermal, and chemical treatments or reinforced externally. For example, work hardening metal like Mn steel and Hadfield steel results in plastic deformation assisted transformation of comparatively soft and tough austenite into hard and brittle martensite, whereas the development of composite coating reinforced with hard microconstituents like hard and stable carbides and borides (such as CrC, CrB, NbC, WC, TiC, TiB) in soft and tough matrix of iron, cobalt, and nickel (Fig. 3.10). These coatings offer better resistance to the external loading, indentation, scratch formation, impact and shocking loading, delamination, thermal softening; all these factors help in enhancing the wear resistance.

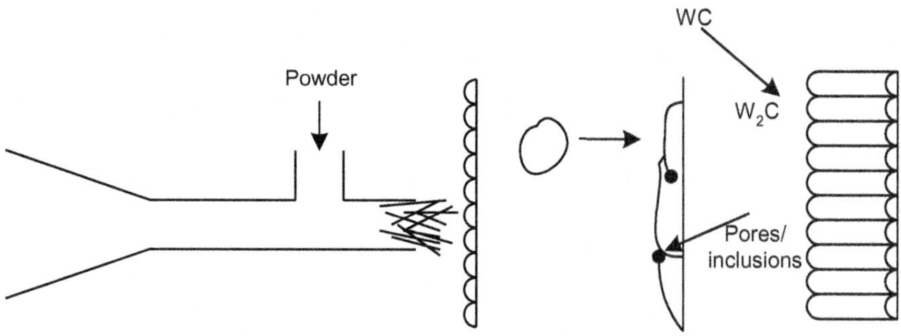

Fig. 3.10 Schematic of thermal spraying and coating

3.4.12 *Thermal Barrier Coating*

Many engineering systems operating at high temperature need protection for survival and for extension of the expected service life. For example, the skin of rockets must be able to withstand extreme temperature condition generated due to frictional heat while entering the earth's atmosphere from space. The surface of such components is protected using suitable coating systems that reduce heat transfer to the substrate. The coating materials acting as barrier for transfer of heat are called thermal barrier coating. Ferritic stainless steels and Cr–Mo steels are expected to provide services at high temperature (about 500–650 °C). The performance of thermal barrier coatings depends on the thermal conductivity of coating material, compatibility of coating material with the substrate in relation to the thermal expansion coefficient determining the ability to withstand thermal cyclic conditions, and thermal stability at high temperature (with respect to the micro-structure and mechanical properties). The application of thermal barrier coating can increase the working temperature up to 200 °C, above the normal operating temperature. Zirconia and yttria-stabilized zirconia are commonly used as thermal barrier coating systems. To deal with the issue of compatibility of thermal expansion coefficient also termed as strain compliance (due to mismatch in base metal and coating material), sometimes another material with thermal expansion coefficient in between the coefficients of substrate and coating materials is utilized, which is called bond coat (Figs. 3.11 and 3.12). The bond coating increases the adhesive strength of coating with substrate. Thermal barrier coatings need to be thin for aerospace applications due to weight consideration, while higher thickness is permitted for other applications such as thermal power plant and automobile engines. Porosity and cracks in thermal barrier coating adversely affect the performance. On the basis of the above observations, the following properties must be possessed by material that can be used for thermal barrier coating:

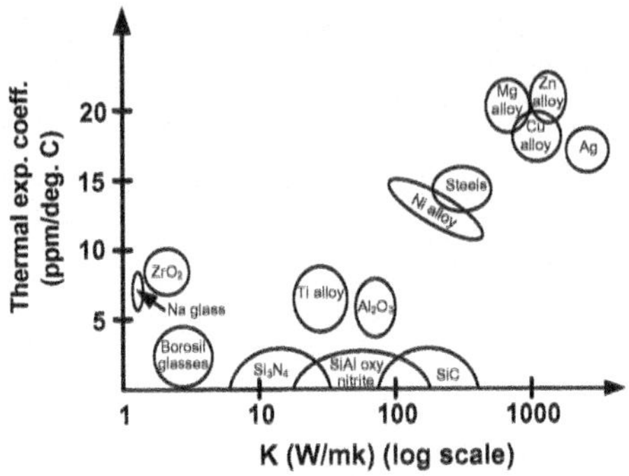

Fig. 3.11 Thermal conductivity and thermal expansion coefficient of common materials

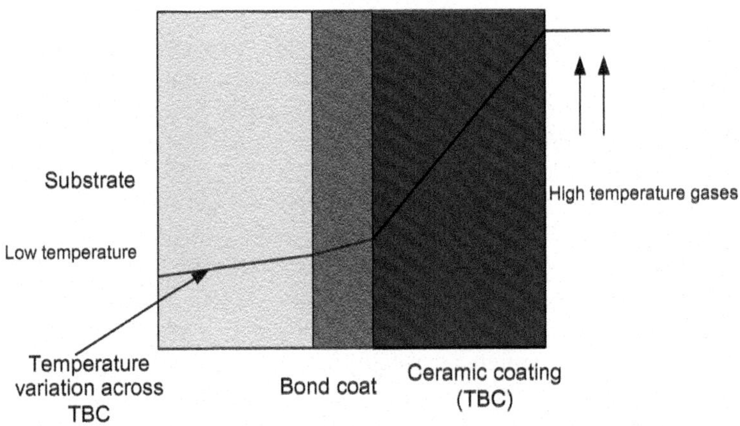

Fig. 3.12 Schematic of thermal barrier coating

- Low conductivity to act as thermal barrier;
- Low density to reduce the possibility of chipping off or spalling owing to high centrifugal forces in rotating components like shafts;
- Lower the difference in thermal expansion coefficient of substrate and coating material so that there is a better strain compliance.

Good chemical and thermal stability of thermal barrier coating are required so that TBCs do not react with service atmosphere (steam and flue gases carrying sulfur) and the substrate at the service temperature.

Thermal efficiency of internal combustion engines such as gasoline vehicles is determined by the extent of heat generated through the combustion of fuel is used for producing power to drive the vehicle. Loss of heat to the coolant (which is used to maintain the cylinder and other components within safe working temperature limit) in turn reduces the thermal efficiency. The application of thermal barrier coatings on the internal surfaces of cylinder and other components will keep them at low temperature while leading to reduction in the loss of heat, so most of the thermal energy of fuel is utilized for producing power to drive the vehicle. This in turn will enhance the thermal efficiency of fuel, or in other words, mileage per unit consumption of the fuel. However, presence of sulfur with hot flue gases of automotive forms H_2SO_4, especially in moist condition. This acid is known to be harmful for ferrous components, as it attacks the iron component in exhaust systems and degrades their life. Sometimes, silicon carbide in thermal barrier coatings reacts with oxygen present in the steam environment to form silicon, which in turn degrades the performance of thermal barrier coatings. Therefore, thermal barrier coatings must be stable in the presence of steam and sulfur.

3.4.13 Functionally Graded Materials (FGM)

Many applications need extremely different property at the surface than that at the core of component, for example, high hardness and wear resistance at surface and high toughness, electrical, and thermal conductivity at interface or at the substrate. The development of coating having continuous gradient in composition, structure, and properties such that a layer of coating material next to the substrate is largely of properties similar to that of the substrate, while a layer of coating material at the top of the surface corresponds to the functional properties desired at the surface. The requirements of such type coatings have led to the development of functionally graded (FG) coatings. Functionally graded coatings are materials having engineered gradient of composition, structures, and/or specific properties (Fig. 3.13). These are developed with the aim of having better property combinations over homogeneous materials of the same composition. However, FGMs under prevailing technological conditions are not economically viable because the complex manufacturing procedure used for the development of FGMs makes them costlier. For example, the development of coating with four different composition and properties needs spraying of coating material one by one; hence, the coating is done by spraying it four times.

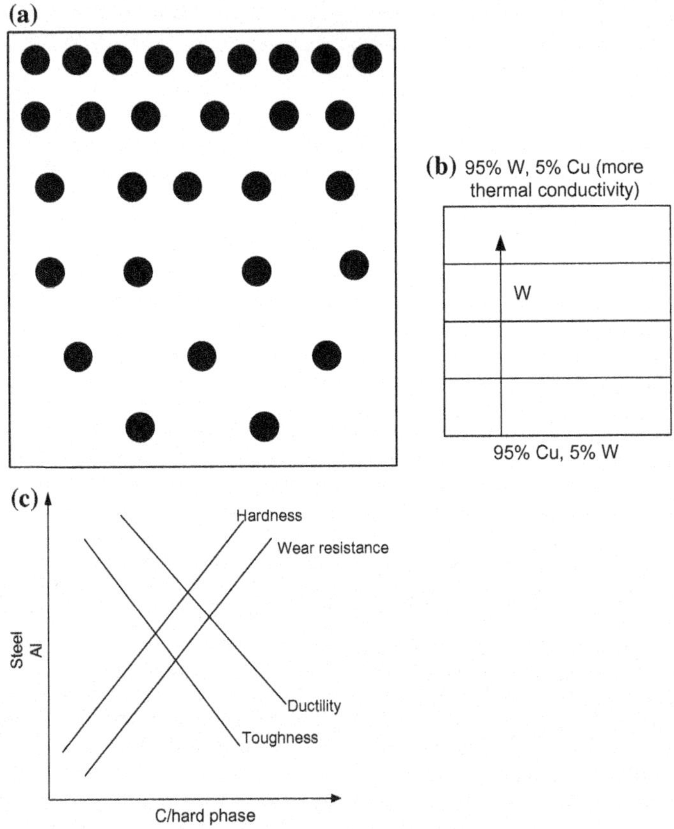

Fig. 3.13 Schematic of functionally graded materials **a** and **b** distribution of element/constituents and **c** variation in properties of steel/aluminum with variation in carbon content/hard constituents

3.5 Common Materials for Surface Modifications for Specific Applications

Materials for surfacing can be divided into eight categories on the basis of composition, hardness, and hardening constituents. These categories have been presented in the following section.

Category 1 belongs to ferrous system of low alloy content having total alloying elements less than 10% and hardness less than 450 HV. These are used for applications needing resistance to impact as butter layer and abrasion resistance under low stress conditions.

Category 2 belongs to ferrous system of low alloy concentration (with higher content Cr) with total alloying elements less than 10% but hardness greater than 450 HV. These are primarily used for hardfacing of components namely blades, scoops used in soils and sand handling equipment.

Category 3 belongs to medium alloy ferrous system having total alloying elements in a range of 10–25%, and hardness of weld surfacing ranges from 450 to 700 HV as per welding process. These are primarily used for abrasive wear resistance application at moderate temperature range from 300 to 600 °C such as weld surfacing of roller of hot steel mills.

Category 4 belongs to medium alloy work hardenable ferrous system with total alloying elements in a range of 10–25% and low hardness in a range of 200–250 HV and subsequently on work hardening during service due to impact loading gets hardened in range of 400–500 HV.

These are primarily used for service conditions involving wear coupled with impact for applications such as crusher and excavator.

Category 5 belongs to high alloyed (Cr–Mn, Cr–Ni, Cr–Ni–Mn) system with total alloying content more than 25% but hardness less than 450 HV. These are used for corrosion resistance application and to develop buttering layer to facilitate welding of incompatible steels.

Category 6 belongs to hardfacing ferrous system having chromium carbides (1500 HV) which primarily comprises Cr and C of ranges 25–30 and 2–6%, respectively, and hardness of surfacing may vary from 500 to 850 HV as per alloying content. These are used for surfacing of equipments used in agricultural, mining equipment, chutes, sandblasting plant, etc.

Category 7 belongs to hardfacing system based on tungsten carbides (2000–2200 HV) embedded in hard matrix of martensite (700–900 HV). These are primarily used for increasing abrasive wear resistance of equipments used for construction and farm such as digging blades used in sandy soil.

Category 8 belongs to hardfacing system based on complex carbides, i.e., more than one type of carbide. Carbides of Cr, V, W, Nb, and Mo are commonly used. Nb-containing system can bear little impact loads. Systems of this category show good resistance to oxidation and thermal softening at 600 °C. These alloys are therefore primarily used for high-temperature wear applications such as pulveriser and sintering plants.

3.6 Common Materials and Their Typical Applications

Wear process	Typical coating options
Abrasion	Al_2O_3, Cr_2O_3, NiCrBSiC, We–Co, TiC–Ni, Cr_2C_2–NiCr, nanosize powders
Cavitation erosion	NiTi, Cu–Ni, NiCrBSiCAlMo, 316 S/S
Liquid impingement erosion	Stellite, Al_2O_3, WC–Ni
Fretting Low amplitude High amplitude	Cr_3C_2–NiCr Cu–Ni, Cu–Ni–In CoMoCrSi, Mo–NiCrBSi, T–800
Abradable	CoNiCrAlY, Ni–graphite, NiCrAl–bentonite

(continued)

(continued)

Wear process	Typical coating options
Impact and sliding	WC–Co, T–800, Cr_3C_2–NiCr, nanosize powders
Galling	Cr_3C_2–NiCr, T–800
Solid particle erosion	WC–Co, Cr_3C_2–NiCr, WC–NiCrBSiC, Cr_3C_2–MCAlY
Adhesive	Al–bronze, Mo–NiCrBSi, Al_2O_3–TiO_2, WC–Co
Scuffing	Mo, Mo–NiCrBSi
Dry sliding	WC–Co, Cr_2O_3, Cr_3C_2–NiCr, NiCrBSi

Questions for self-assessment

1. What are metal properties need to be considered for selection of materials for wear resistance applications under following conditions?

    a. Adhesive wear
    b. Abrasive wear
    c. Erosive wear
    d. Cavitation
    e. Fretting wear

2. What is significance of thermal stability and affinity of metals with gases with regard to adhesive wear?
3. How do metal properties affect the galling resistance?
4. How does microstructure of material affects the wear resistance?
5. Describe response of following metals in respect of wear

    a. Mild steel
    b. Alloys steel
    c. Stainless steels
    d. Hadfield steel
    e. White iron
    f. Ni alloys
    g. Cobalt alloy
    h. Aluminum alloys

6. Schematically show general guidelines for selection of metals for wear resistance application.
7. Write about functionally grades metal and their significance using schematic diagram
8. What for thermal barrier coatings are used?
9. Describe fundamentals related with thermal barrier coating regarding issues and applications.
10. Describe various wear-resistant metals available for surface engineering.

Chapter 4
Surface Engineering by Changing the Surface Metallurgy

4.1 Approach

The approach of modifying the surface characteristics by just changing the surface metallurgy without varying the chemical composition is primarily based on the application of three types of material strengthening mechanisms: (a) grain refinement, (b) transformation hardening, and (c) strain hardening. Surface modification techniques based on the above three mechanisms have been developed over the years. Generally, the first two strengthening mechanisms—grain refinement and transformation hardening—jointly contribute in strengthening the surface layers in all those methods (flame, induction, and laser hardening, TIG and laser re-melting), where surface modification is based on the use of thermal energy, whereas strain hardening and plastic deformation assisted transformation result in the desired hardening for those processes (burnishing, shot peening, friction stir processing) which employ mechanical energy for surface modification.

It is important to note that the relative effectiveness of each of the processes varies significantly with the metal system in consideration. Thermal energy based methods are generally more effective with those metal systems (medium and high carbon steel, alloy steel, cast iron) that form hard microconstituents (like bainite and martensite in steels) owing to the thermal cycle (heating–holding–cooling) experienced by them during processing of transformation hardening metals than those metals that are subjected to grain refinement only during processing.

Transformation hardening-based methods (mostly use thermal energy), namely flame, induction, and laser hardening, have been used with ferrous metals having carbon content more than 0.3% to ensure the development of hard microconstituents during processing. The medium, high carbon steel, alloy steel, and cast iron offer high hardenability because of the low critical cooling rate that is necessary for the formation of hard microconstituents like martensite; therefore, surface hardening of these metals can be done easily using a sequence of rapid heating and cooling. However, surface hardening of work hardening (non-heat treatable or

D. K. Dwivedi, *Surface Engineering*,
https://doi.org/10.1007/978-81-322-3779-2_4

non-transformation hardenable) metals like low carbon steel and austenitic stainless steel, Mn steel, cobalt base alloys, non-heat treatable aluminum alloys can be carried out using either grain refinement approach (melting followed by rapid cooling) or work hardening approach. The work hardening-based methods—shot peening, friction stir processing, and shot blasting—are more effective for low stacking fault energy metals (gold, copper, Mn steel, had field steel, cobalt alloy) than high stacking fault energy metals (Mg and Al).

To overcome this problem, two different kinds of case hardening are used: one for those steels having carbon less than 0.3% and another for those having carbon more than 0.5%. The combination of properties (hardness and toughness) can be obtained only by controlling the microstructure across the product section. We know that high hardness in steels can be obtained by having the martensitic structure, whereas for toughness in steel, we should have comparatively soft phases such as ferrite, bainite. Case hardening is done to achieve the martensitic structure or nitrides at the surface and soft phases in the core. It is easier to obtain the martensite in high carbon steel ($C > 0.5\%$) than the low carbon steel ($C < 0.3\%$). Hence, case hardening of the low carbon steel first needs the addition of carbon to produce martensite of high hardness during subsequent hardening treatment. Case hardening of high carbon steel is done directly by shallow hardening without the addition of carbon.

4.2 Transformation Hardening Methods

The development of a hardened case using **transformation hardening** methods is mainly applied for high carbon steel, and it is based on shallow hardening. However, laser and electron beam hardening can also be used for low carbon steel through grain refinement and favorable phase transformations. The shallow hardening is based on the principle of martensitic transformation, which is carried out by controlled heating of surface layer to the austenitic temperature ranging from 800 to 850 °C (at about 50° above the upper critical temperature of the steel depending upon the alloying elements, e.g., carbon content in the steel), followed by rapid cooling to get martensitic structure. Depending on the method of heating adopted for austenitizing, transformation hardening methods are classified as (a) flame hardening, (b) laser hardening, and (c) induction hardening. Heat treatment of transformation-hardened surfaces of steel alters the microstructure and hardness (Fig. 4.1). Annealing of case-hardened surfaces causes more softening than tempering as annealing results in coarse ferrite and pearlite formation, while tempering causes softening of martensite by diffusing out some carbon from supersaturated solid solution of carbon in iron, i.e., martensite.

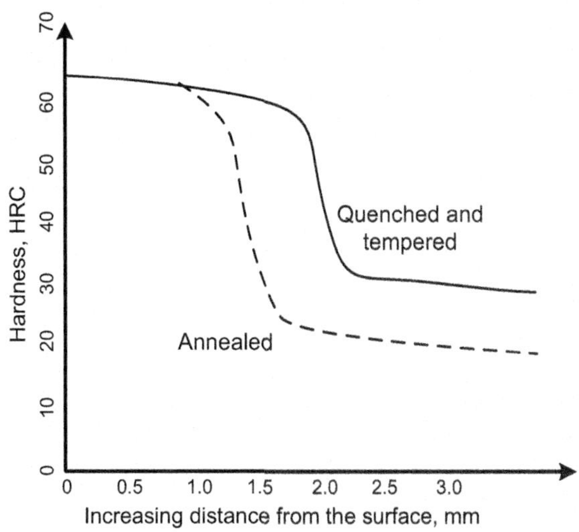

Fig. 4.1 Hardness variation of a high carbon steel as the function of distance from the surface subjected has different heat treatments

4.2.1 Flame Hardening

This process uses oxy-fuel flame for heating the surface of the component to be hardened, and once the heating up to the desired temperature is achieved for sufficient duration, it is subsequently quenched by a water jet following the heating torch (Fig. 4.2). The speed and relative position of heating torch with respect to water jet nozzle (distance between the flame and water jet) determine the holding time above the austenitizing temperature at which the quenching is performed to realize the martensitic transformation. These two factors in turn affect the depth of the case developed on the components and hardness distribution at the surface and subsurface regions. The depth of case is measured in terms of thickness of the subsurface zone up to which the desired level of the hardness is achieved.

4.2.2 Induction Hardening

Induction hardening is based on the principle of rapid heating (using AC electric current induced at the surface and subsurface regions) followed by quenching of the component that is to be hardened (Fig. 4.3). The current is induced by bringing the surface of the component closer to a coil through which high-frequency current is passed for a very short period. The flow of current is limited to very thin layer of component due to skin effect and proximity effect caused by high-frequency AC current. Electrical resistance heating causes rise in temperature in range from 300 to 1200 °C in thin layer through which induced current flows. To achieve proper austenitization up to desired depth, heating temperature and soaking time (thermal cycle, i.e., heating, soaking, and cooling steps) must be closely controlled.

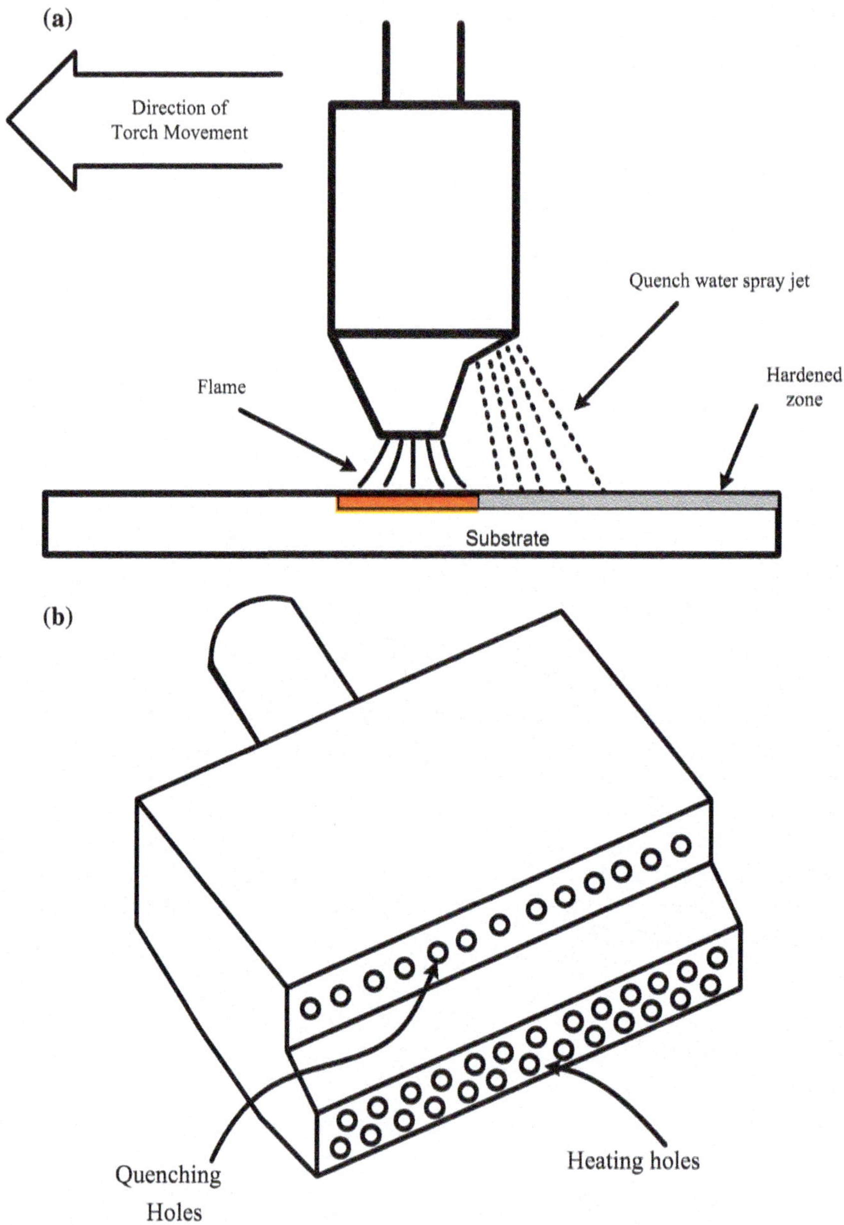

Fig. 4.2 Schematic of flame hardening, **a** arrangement of different components and **b** flame hardening torch

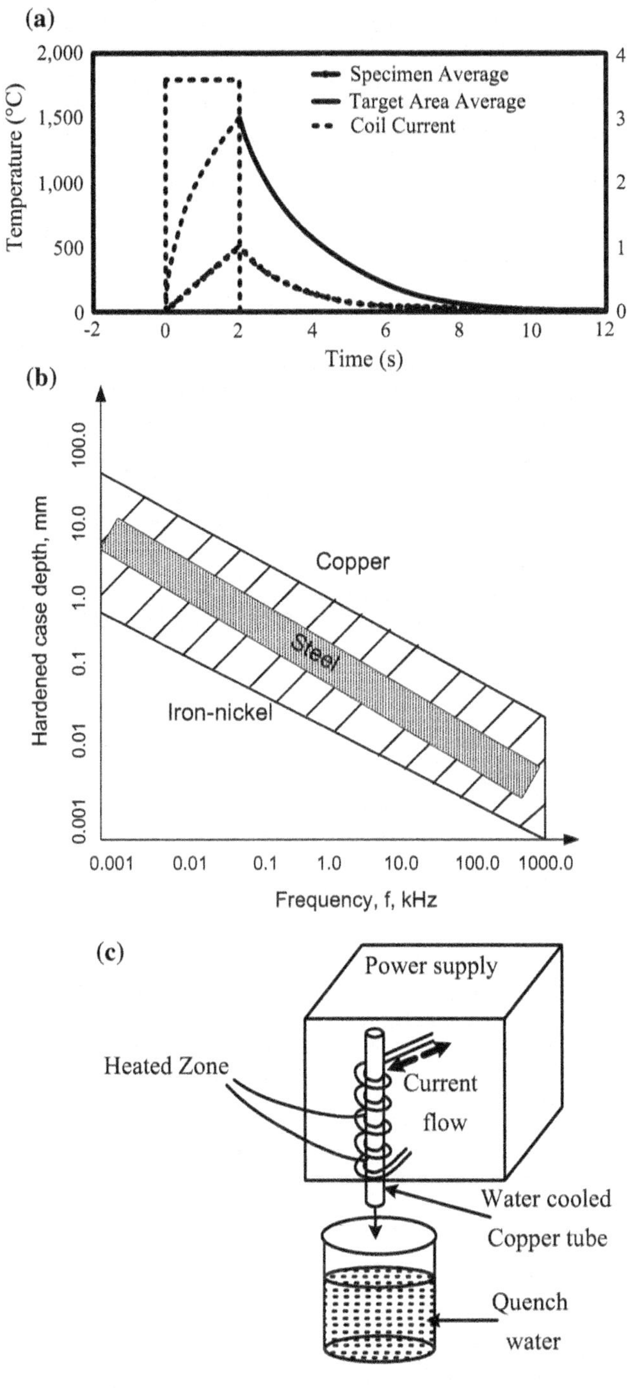

Fig. 4.3 Induction hardening, **a** variation in temperature as function of time, **b** effect of current frequency on case depth, and **c** schematic arrangement for induction hardening

Depending on the frequency of current through the primary coil, eddy current is induced to a certain depth, which in turn causes rapid heating (by electrical resistance heating) of the material within a very short time span. Depth up to which current is induced determines the depth of case hardening (Fig. 4.2b). The penetration depth (t in m) of induced current depends on frequency (f in Hz), specific permeability (μ), and electrical conductivity of job (e in mho/m).

$$\text{Penetration depth (m)} = 1/(\pi f \mu e)^{0.5}$$

Heated surface is then quenched for necessary martensitic transformation. Schematic arrangement of induction hardening for cylindrical rod is shown in Fig. 4.2c. Depending upon the section thickness, quenching can be realized in two ways: one is jet of water and the other is just switching off the current that passes through the induction coil. Thus the bulk substrate material below the case causes rapid cooling. The depth of case is determined by the extent up to which the subsurface region is heated above the upper critical temperature (as per the frequency of current through the primary coil) and the case experiences the transformation hardening. Case depth and frequency are inversely related.

4.2.3 *Laser Beam Hardening*

Laser can be used for improving the surface properties using three primary approaches: (a) heating for transformation hardening, (b) melting for controlled alloying, grain refinement, and cladding, and (c) shock peening to develop compressive residual stress (Fig. 4.4). In laser hardening process, a laser (light amplification and simulated emission of radiation) of high energy density is used for rapid heating or austenitizing of the surface layer followed by self-quenching (because beneath the base metal at room temperature acts heat sink), as the beam is moved away from the heated zone. A proper combination of laser power density

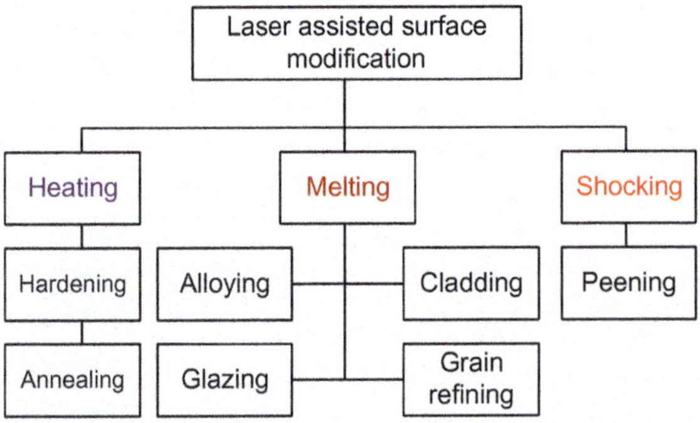

Fig. 4.4 Block diagram showing mechanisms and process for laser-assisted surface modification

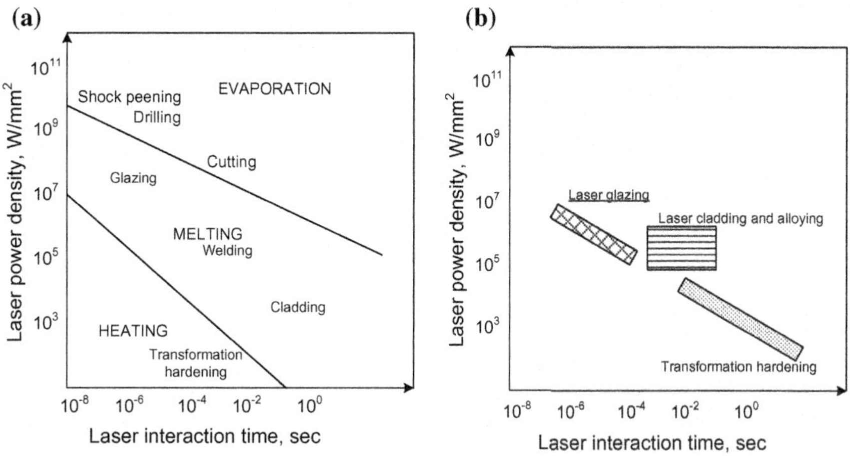

Fig. 4.5 Laser power density and interaction time relationship for, **a** a large range of processes and **b** laser cladding and hardening processes

and laser-substrate interaction time is crucial for austenitization up to the desired depth for the case hardening purpose (Fig. 4.5a, b). Cooling by underlying low-temperature core material causes the necessary transformation of austenite into martensite for shallow hardening.

It causes hardening of the outer layer of the workpiece typically up to a depth varying from 0.1 to 1.5 mm. High depth of hardening needs a compromise between heating of large volume of metals and fast enough heat dissipation needed for requisite martensitic transformation. A proper combination of laser power, beam diameter, and laser scanning speed results in requisite hardness up to the desired depth. Type of laser such as continuous wave, pulse wave laser (nano, peco, femto second laser), determines the liquid zone geometry and heat-affected zone (Fig. 4.6).

High energy associated with laser leads to austenitizing of the thin layer below the steel surface, which in turn determines the case depth (Fig. 4.7). Section size of the component determines the quenching rate after austenitizing. The microstructural modification and hardness variation as function of depth from the surface of laser hardened of cast iron is shown in Fig. 4.8a, b. Therefore, laser hardening is found suitable for case hardening of heavy section medium and high carbon steel components.

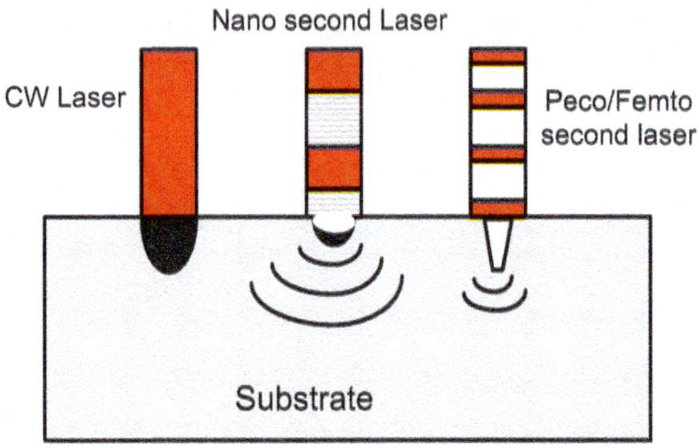

Fig. 4.6 Schematic representation of effect of continuous and pulse lasers on substrate and heat-affected zone

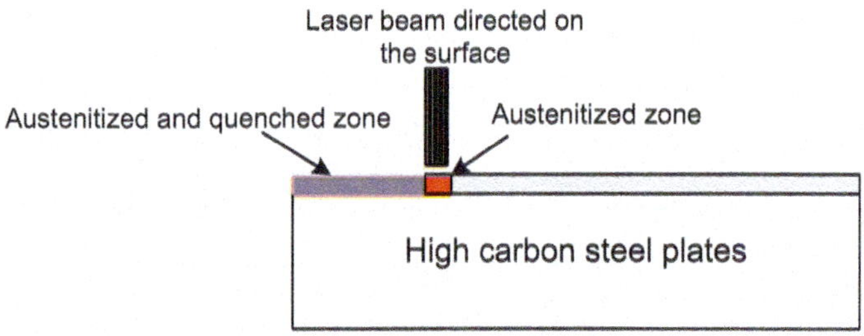

Fig. 4.7 Schematic of laser hardening processes

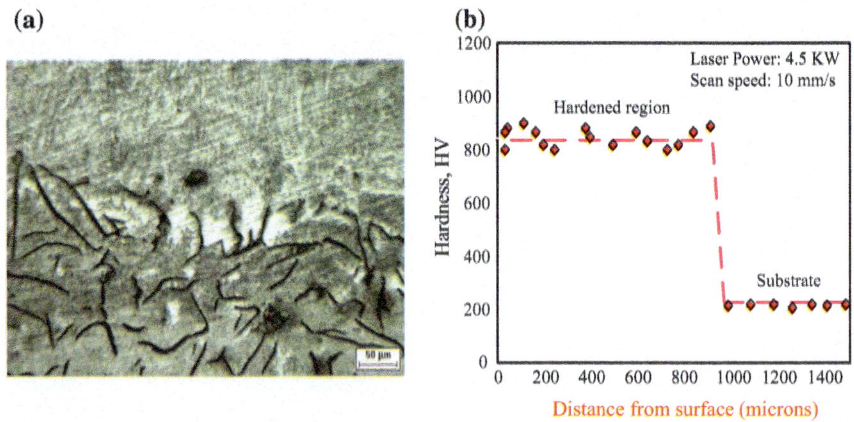

Fig. 4.8 Laser hardening, **a** microphotograph laser hardening gray cast iron and **b** hardness variation as a function of distance from the surface (Sharif 2007)

Typical applications of laser hardening are given below.

Component	Material
Axle bearing seat	AISI 1035 steel
Blanking die	Tool steel
Engine valve	Alloy steel
Gear teeth	Steel
Shaft	Steel
Piston ring	Steel
Steering gear housing	Malleable cast iron
Typewriter interposer	AISI 1035 steel
Cutting edge	Steel
Gear teeth	AISI 1060/low alloy steel
Mandrel	Martensitic stainless steel
Press brake tools	Steel
Tool bed	Cast iron
Turbine blade edge	Gray cast iron
Diesel engine cylinder	Cast iron

4.3 Re-melting of Base Metal or Modified Surfaces Using Laser and TIG

Re-melting can appreciably lead to the improvement in the mechanical and tribological performance of component or bulk metals that are developed with coarse-grained structure by using processes like sand mold casting, investment casting processes, and coatings or overlays deposited by means of low energy density processes such as flame spraying, gas welding surfacing, shielded metal arc welding. Schematic of re-melting of flame-sprayed coating is shown in Fig. 4.9. The re-melting of a thin layer metal at the surface using comparatively high energy density processes like laser and electron beam, plasma, and gas tungsten arc followed by rapid solidification results in extremely fine grain structure and sometimes allows favorable transformation during the hardening of metals like cast iron, medium and high carbon steels (Fig. 4.10a). Refinement of the grain structure therefore improves the hardness of the coating (Fig. 4.10b). Similarly this approach can also be used to improve the surface properties of any metallic system (such as components made by casting) have coarse grain structure by grain refinement. High cooling rate during solidification results in significant refinement of grain structure, and subsequently, high cooling rate (cooling from high temperature to room temperature) may cause transformation hardening. In the case of non-transformation hardening metals like Al–Si alloy and low carbon steel castings, the refinement of grain structure helps in improving the mechanical and tribological performance, while in the case of transformation hardening metals like cast irons, high carbon,

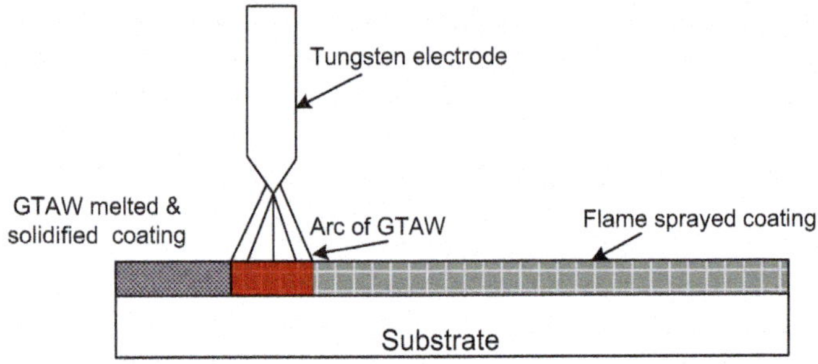

Fig. 4.9 Schematic of re-melting of the flame-sprayed coating by TIG arc welding

and alloy steel; metallurgical transformation plays a more important role as compared to grain refinement (Fig. 4.9).

4.4 Plastic Deformation-Based Approaches

The changing of the surface metallurgy using mechanical energy for the improvement of the surface properties is primarily based on the principle of controlled deformation of surface layers that occurs through various processes such as burnishing, shot peening, and friction stir processing. The controlled plastic deformation of metals near surface layers results in the work hardening effect by developing large number of dislocation (Fig. 4.11). Dislocations produce strain fields around themselves. These strain fields interfere with the movement of other nearby dislocations under the influence of external load. Any resistance to the movement of dislocations increases the stresses required to cause deformation, which means yield strength improves with the development of dislocations. However, the disadvantage of methods based on this principle is that the considerable reduction in ductility and increase in ductile to brittle transition temperature (DBTT) in strain-hardened materials. Further, the extent of hardening for the given amount of deformation depends on stacking fault energy (SFE) of material under consideration. Metals (like Mn steel, Austenitic stainless steel, Hadfield steel, gold, Cu) of low SFE are strained hardened appreciably.

Further, localized surface deformation also increases the capability of carrying tensile load and fatigue strength by developing the residual compressive stress. The localized surface deformation using the above-mentioned methods for improving the fatigue performance is based on the simple concept of lowering the effective applied tensile stresses through the induction of residual compressive stress. The compressive stress to some extent neutralizes or lowers the magnitude of externally applied tensile stress. Therefore, this approach is found effective only when fatigue load is tensile in nature and magnitude is lower than yield strength. Moreover, this approach has marginal effects on the fatigue performance under low cycle fatigue

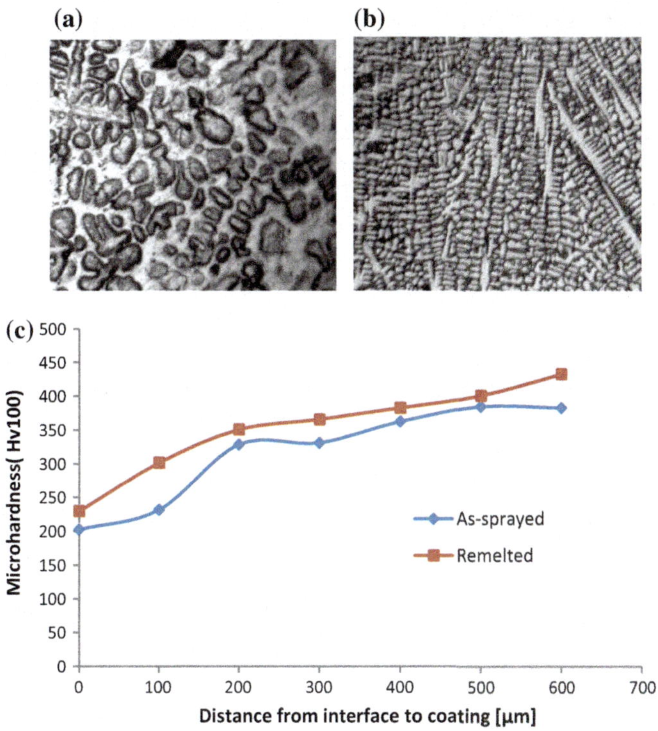

Fig. 4.10 Micrographs of flame-sprayed coating of Ni–Cr–B–Si alloy, **a** in as-sprayed conditions, **b** TIG re-melted (200X), and **c** microhardness distribution before and after re-melting (Dhakar 2010)

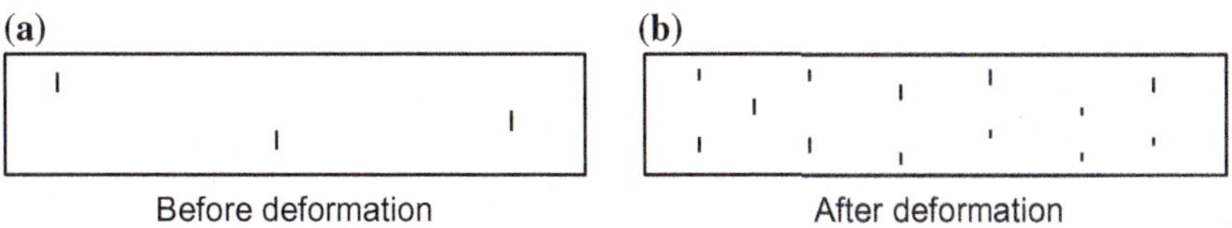

Fig. 4.11 Schematic of dislocations in metal system, **a** before and **b** after plastic deformation

conditions when the fluctuating loads and corresponding stresses are more than yield strength. The improvement in fatigue performance by this approach can vary from 20 to 30%. There are many methods such as shot peening and contour rolling, which can be used for inducing compressive residual stress. All these methods are based on the principles of differential dimension or volumetric change between surface layer and sub-surface of the substrate through the application of localized stresses at the surface. The reduction in tensile stresses or development of compressive residual stresses also leads to resistance to stress corrosion cracking (SCC).

4.4.1 Shot Peening

In case of shot peening, high-speed steel balls are directed toward the surface of the substrate on which the compressive residual stress is to be developed. Steel balls of 5–10 mm diameter are accelerated to high velocity of about 100 m/s and shot at the surface to be treated (Fig. 4.12a). The impact of shots produces indentation through localized plastic deformation at the surface layers, while metal layers below the plastically deformed surface layers are subjected to elastic deformation (Fig. 4.12b). Material further deeper from the surface is unaffected by shots and plastic deformation occurring at the surface. Elastically deformed layers tend to regain their dimension, while plastically elongated surface layers resist any comeback. Since both plastically and elastically elongated layers are metallurgically bonded together, the

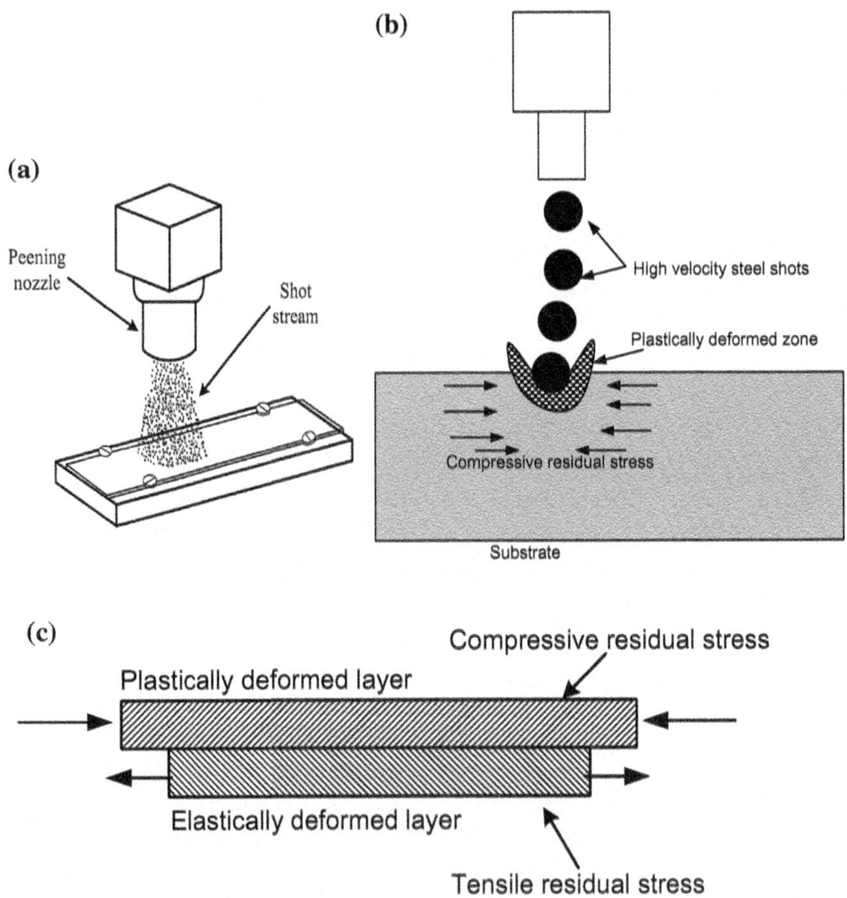

Fig. 4.12 Shot peening process showing schematic of, **a** process, **b** deformation due to impact of shot, and **c** elastic, plastic deformation and residual stress

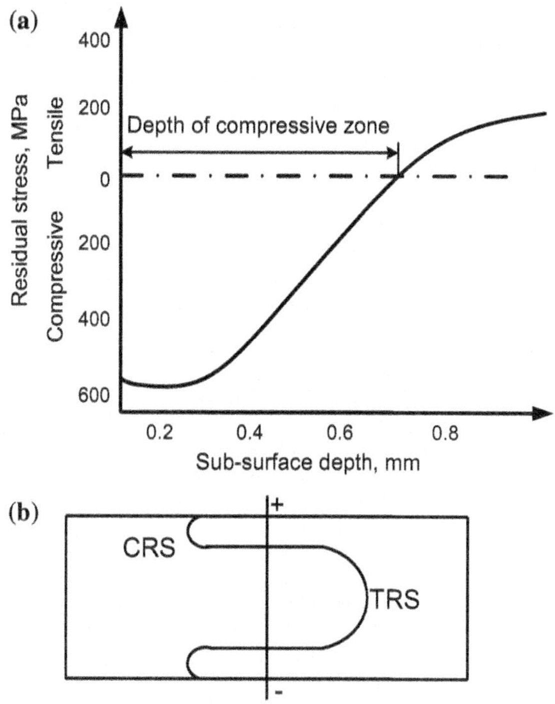

Fig. 4.13 Schematic of development of residual stress, **a** increasing depth from the surface and **b** residual stress pattern in cylindrical component after shot peening

elastically elongated undersurface metal layer tends to put plastically elongated surface layer under compression; however, the elastically elongated undersurface layer comes under tension (Fig. 4.12b, c). Thus, the residual compressive stresses are induced at shot-peened surface. The presence of tensile residual stress below the surface is not considered to be much damaging for fatigue life, as mostly the fatigue failures commence from the surface (Fig. 4.13a, b).

Metal immediately below the surface is plastically deformed, and those layers of metal further below the plastically deformed zone, develop an elastic deformation zone. The elastically deformed zone tries to come back to initial state, but the same is resisted by the plastically deformed layer. This situation leads to the development of residual compressive stress at the surface and tensile stress in core of shot-peened metal which results in the balancing of the stress system (Fig. 4.13).

4.4.2 *Burnishing and Contour Rolling*

Burnishing and contour rolling are similar types of processes, where the localized surface deformation is achieved by pressing a tool against the rotating or sliding surface of the workpiece to be processed (Fig. 4.14a, b). However, the tool used for burnishing and contour rolling differs significantly. The burnishing tool is made of

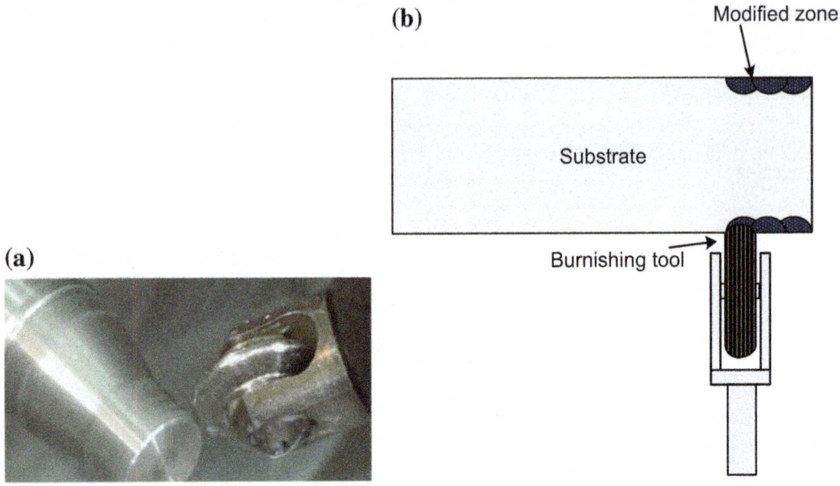

Fig. 4.14 Burnishing process, **a** photograph of burnishing tool and **b** schematic of burnishing process

hardened steel with rounded tip, while contour rolling mainly comprises a rotating disk with rounded edge or circumference as shown in Fig. 4.14a.

The application of normal load onto the surface through ball generates high hertzian compressive stresses in the material. Three dimensional stress state developed at the contacting surface results in plastic deformation above yield stress (Fig. 4.15a–d). Burnishing flattens asperities present on the surface in the form of peaks and valley, thereby improving the surface finish (Fig. 4.15c–d). The influence of contour rolling is similar to that of shot peening. Near-surface layers are subjected to severe plastic deformation and work hardening, which increases the hardness besides developing the residual compressive stresses (Fig. 4.16).

4.4.3 Friction Stir Processing

Friction stir processing (FSP) is an effective solid-state surface modification technique for the enhancement of the mechanical and tribological resistance of metallic systems like aluminium, titanium using two approaches, which are (a) structural modification of near-surface layers without any change in composition and (b) structural modification of near-surface layers coupled with the change in composition such as composite layers (Fig. 4.17). Typical multi-pass friction stir processed (FSPed) surface of AA7039 alloy is shown Fig. 4.17b. Furthermore, FSP is a viable means of producing localized composite zones in cast Al–Si alloy components. Such improvements have vital implications for the manufactured components

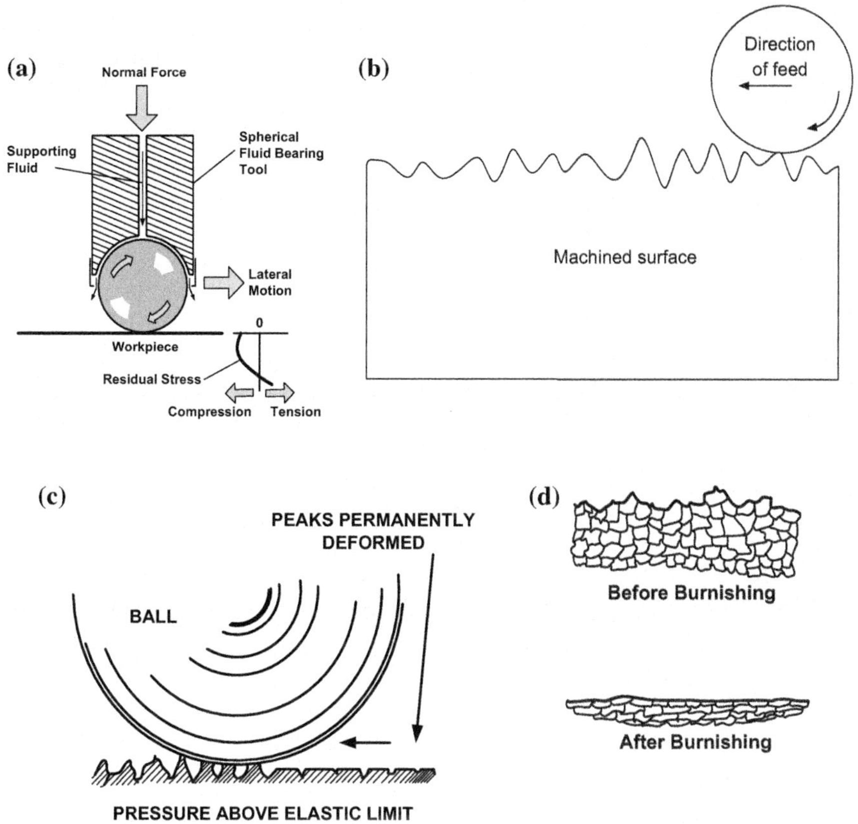

Fig. 4.15 Burnishing peening, **a** schematic of tool for burnishing on the surface, **b–d** deformation of surface and near-surface layers due to burnishing

of diesel engines and for critical and high integrity components. The wear performance of a component is predominantly governed by surface characteristics such as hardness and fine grain structure. In general, an increase in hardness and grain refinement improves the wear resistance (Fig. 4.18). Similarly, structural characteristics (size and distribution of microcontituents) and severe strain induced during FSP of near surface layers can also improve wear behavior. The modification in mechanical characteristics (hardness and fatigue resistance) and metallurgical characteristics (grain size, precipitate characteristics and their distribution) of FSP surfaces therefore affects sliding wear resistance and mechanical properties (Fig. 4.19). Grain refinement after FSP increases the elongation significantly.

The FSP generates three distinct microstructural zones: the nugget, the thermomechanically affected zone (TMAZ), and the heat-affected zone (HAZ). The nugget is the region through which the tool piece pin passes and experiences large deformation and high temperature. It generally consists of fine equiaxed grains due

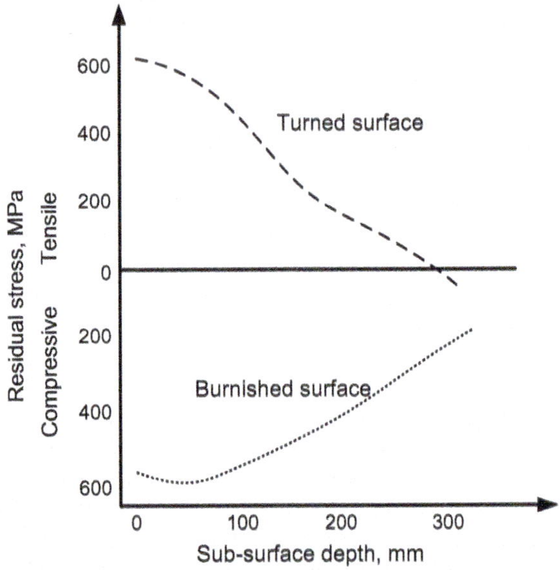

Fig. 4.16 Effect of burnishing on residual stress pattern of the machined component

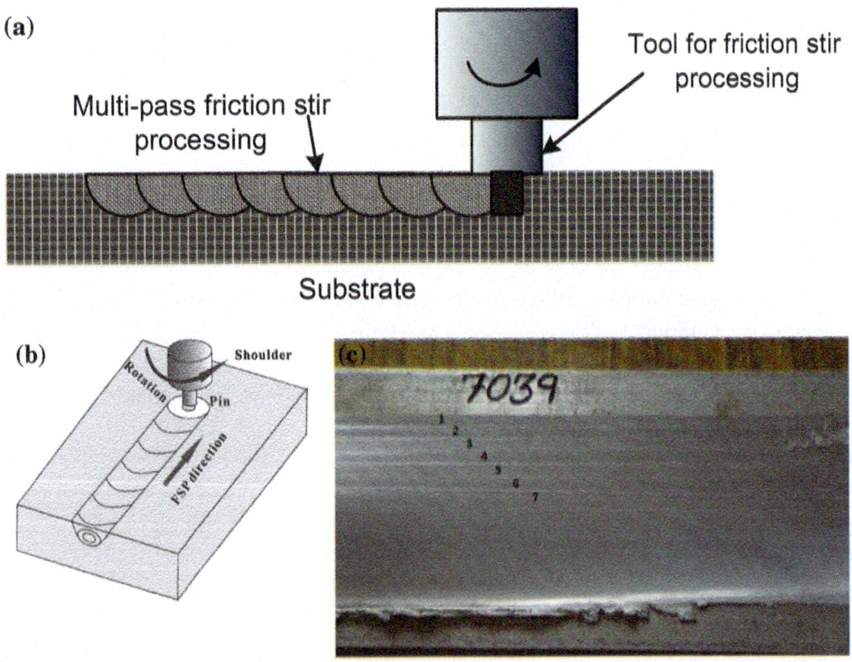

Fig. 4.17 Friction stir processing, **a**, **b** schematic of friction stir process and **c** typical multi-pass FSPed surface

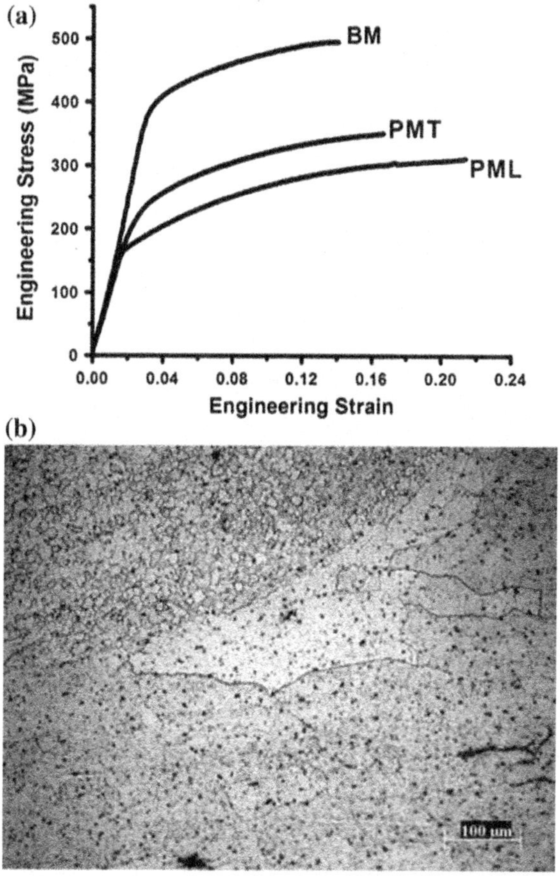

Fig. 4.18 Effect of friction stir processing of AA 7039 Al alloy on **a** tensile properties of specimen from longitudinal (PML), transverse (PMT) direction, base metal (BM) and **b** microstructure

to recrystallization (Fig. 4.1). The TMAZ adjacent to the nugget is the region where the metal is plastically deformed as well as heated to a temperature, which is not sufficient to cause recrystallisation. The HAZ experiences only the heating effect, and there is no plastic deformation. The tool is moved relative to substrate surface such that the tool traverses, until the entire selected area is friction stir processed. Friction stir processing refines grain size due to recrystallization caused by plastic deformation coupled with heat generation well above recrystallization temperature. As the friction processed zone cools, it results in a defect-free recrystallized fine grain microstructure.

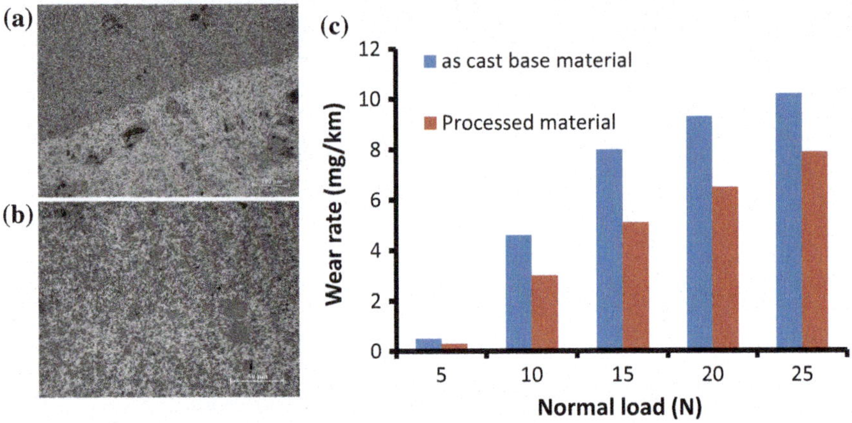

Fig. 4.19 Effect of friction stir processing on, **a**, **b** microstructure of cast Al–17Si–0.5 Mg alloy and **c** wear rate of alloy in as cast condition

Questions for self-assessment

1. What are metal strengthening mechanisms?
2. List the surface modification processes with corresponding metal strengthening mechanism.
3. What is prerequisite for hardening for laser hardening?
4. Explain the mechanism of transformation hardening.
5. Why does the quench steel with limited hardenability show hardness variation across the section.
6. What should be done to change the depth of hardening in (a) induction hardening and (b) flame hardening?
7. How do laser power and laser interaction time affect heating for various surface engineering treatment?
8. What are common applications of laser hardening processes?
9. Distinguish the laser alloying and laser cladding in respect of dilution.
10. How does laser hardening affect the structure and properties of gray cast iron?
11. How can the laser and TIG arc be used for improving the characteristics of the coatings and surfaces?
12. How does plastic deformation affect the properties of metals?
13. Describe underlying principle of (a) burnishing, (b) shot peening, and (c) friction stir processing.
14. Why shot peening and burnishing improve the fatigue resistance.
15. How does friction stir processing of cast Al–Si alloy affect the mechanical and wear properties?

Chapter 5
Surface Engineering by Changing the Composition

5.1 Approach

The surface engineering of components by changing the chemical composition of surface and near-surface layers is primarily conducted to increase surface hardness, residual compressive stresses and other properties as per need. These factors in turn enhance the tensile and fatigue performance, reduce stress corrosion cracking tendency, and increase the wear resistance. The change in surface composition of the component is generally achieved without any major dimensional variation through the application of two approaches: (a) diffusion and (b) controlled melting of the near-surface layers with alloying elements. The selection of elements (carbon, nitrogen, carbon, boron, etc.) to be used for the modification of the surfaces depends on the purpose of surface modification (increasing wear resistance, mechanical performance), size and shape of the components determining the complexity of geometry affecting the distortion tendencies and possibility of adverse effect on the properties of the core of component due to high temperature exposure for longer duration. Schematic diagram shows range of hardness which can be achieved by chemical changes (through diffusion) using different processes (Fig. 5.1). It can be observed that boronizing results in significantly higher hardness than carburizing and nitriding. Thus, choice of the case hardening process by modification of chemistry of the near-surface layers is influenced by maximum achievable hardness besides other factors.

D. K. Dwivedi, *Surface Engineering*,
https://doi.org/10.1007/978-81-322-3779-2_5

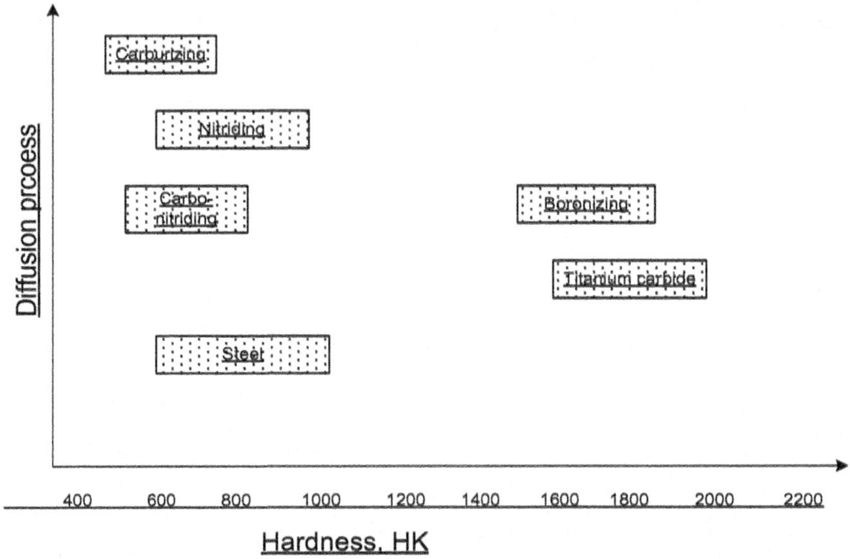

Fig. 5.1 Range of hardness achieved after case hardening by different diffusion processes

S. No.	Elements used	Modification processes	Mechanism
1	Carbon	Carburizing and plasma carburizing	Diffusion
2	Nitrogen	Nitriding and plasma nitriding	Diffusion
3	Carbon and nitrogen both	Cyaniding, carbo-nitriding, nitro-carburizing	Diffusion
4	Boron	Boronizing	Diffusion
5	Cr, Ni, etc.	Laser, plasma, TIG alloying	Melting
6	Ions	Ion implantation	Damaging crystalline structure

The modification of chemical composition of the surface to be engineered is designed to achieve the desired objectives through suitable strengthening mechanism such as solid solution strengthening, precipitation hardening, grain refinement, and transformation hardening. Some of the processes such as carburizing and carbo-nitriding after changing the chemical composition need additional steps of hardening the component through transformation hardening. Low-temperature processes such as nitriding performed in ferritic zone helps to avoid (a) additional steps for transformation hardening, (b) distortion tendency, and (c) deterioration in mechanical properties of the core component.

The surface modification processes involving the enrichment of surface and near-surface layers by carbon and nitrogen such as carburizing, carbo-nitriding, and cyaniding are mainly used for low carbon steel components. The addition of carbon facilitates the formation of hard martensitic structure during subsequently

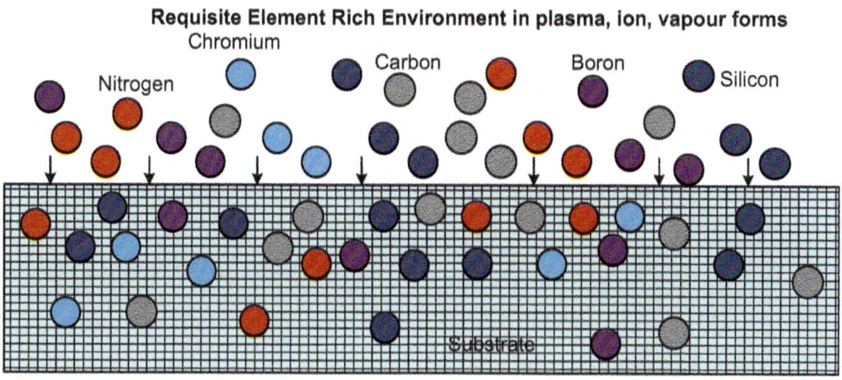

Fig. 5.2 Schematic of surface chemistry modification by diffusion-based approaches

quenching heat treatment cycle which included controlled heating to the austenitic temperature range followed by rapid cooling. In some of the methods, such as nitriding and carbo-nitriding, nitrogen is added to form hard nitrides near the surface. Similarly, boron, chromium and aluminum are introduced during boronizing, chromizing, and aluminizing processes, respectively. Basic approach of surface chemistry modification, for case hardening is shown in Fig. 5.2.

Carburizing, nitriding, and shallow hardening processes increase the performance of tensile and fatigue load carrying capability besides improving wear resistance primarily due to two reasons: (a) increasing surface hardness and (b) developing residual compressive stresses (Fig. 5.3). The development of compressive residual stress is mainly associated with the development of phases during quenching (of carburized samples) and nitriding having higher specific volumes than those are present before processing. The presence of residual compressive stresses negates the effect of external tensile stresses which in turn decreases the crack initiation and propagation tendency and hence increases the tensile and fatigue performance. The increase in hardness delays the crack nucleation stage which in turn increases the fatigue life.

5.2 Carburizing

This method is based on simple approach of developing carbon rich surface and subsurface layer through diffusion of carbon up to certain depth, especially in low carbon steel under favorable conditions; these conditions are: (a) active carbon rich environment which is generally developed through thermal dissociation, (b) high temperature (850–950 °C) which is enough to facilitate diffusion and achieve austenitic state, so that the component has greater acceptability owing to high solid state solubility of carbon in iron, and (c) increase in carbon concentration up to a desired depth is caused by holding the component for a long enough duration

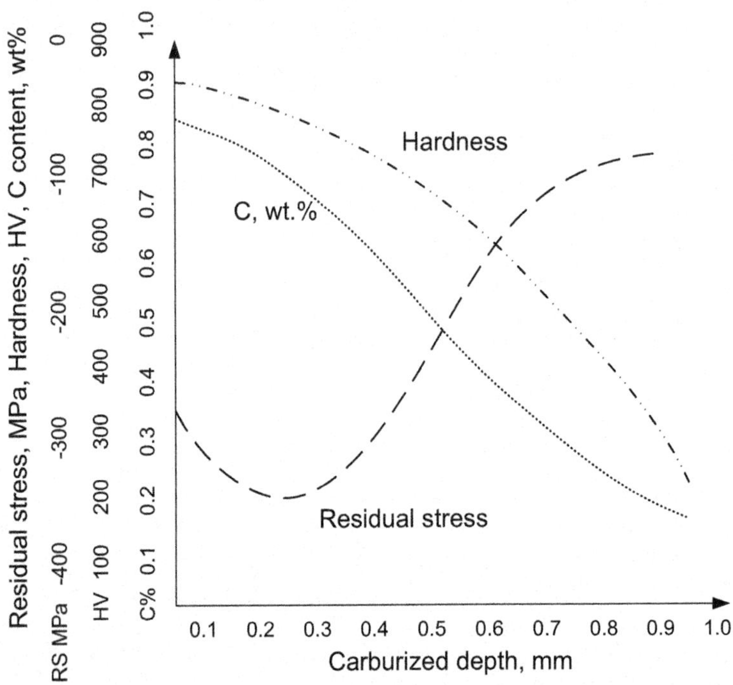

Fig. 5.3 Variation in carbon concentration, hardness, and residual stress distribution increases in depth from the surface after carburizing

(Fig. 5.4a). High temperature increases diffusion rate which in turn reduces the time required for penetrating the carbon up to a desired depth. At a given temperature, the depth of carbon enrichment varies with square root of time (Fig. 5.4b). Usually carbon concentration increases from 0.2 to 0.8 wt%, as further higher carbon addition increases the amount of retained austenite after conducting the quenching treatment because of the reduction in M_f (martensite finish temperature) below the subzero temperature condition. General thermal cycle used for carburizing process is shown in Fig. 5.4c. Gradient in the concentration of the carbon content always exits (maximum at the surface then gradually decreases in the subsurface regions). Therefore, in an effort (increasing carburizing temperature and time) to get high carbon (about 0.5–0.8%) in subsurface region (up to desired), carbon concentration at the surface if crosses the optimum limit leads to the development of softer zone at the surface owing to increase in the amount of retained austenite and maximum hardness zone is shifted to the subsurface regions.

$$2CO \rightarrow C + CO_2$$

According to the kind of medium or environment used for carburizing, the process is termed as solid, liquid, gas, and vacuum carburizing. It is important to

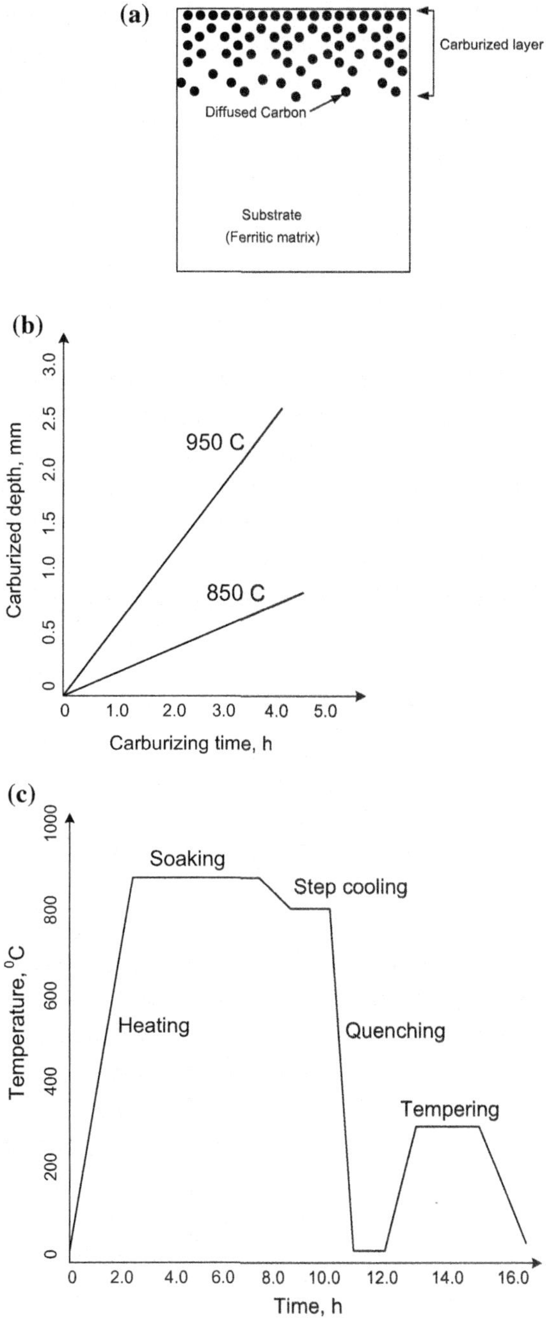

Fig. 5.4 Schematic showing carburizing process
a principle, **b** effect of time and temperature on depth of carburized layer, and
c thermal cycle of carburizing process

note that carburizing increases carbon content only in the component; subsequently, the heat treatment is necessary for producing hard martensite for achieving the desired hardness. Carburizing after hardening treatment results in increase in specific volume which in turn introduces residual compressive stress at the surface.

5.2.1 Solid Carburizing

It uses the mixture of charcoal and barium carbonate (energizer) to develop active carbon rich environment in a heated chamber. Component to be carburized is packed in a closed chamber along with the charge mixture (Fig. 5.5). High temperature results in the release of active carbon through the chemical reactions that are necessary for the diffusion to take place through the surface. There are two main limitations of the method: (a) lack of proper control over the temperature and the reactions taking place inside the chamber, thereby leading to a poor control over the case depth, and (b) carburized component cannot be quenched directly from the carburizing temperatures due to the inherent processes of nature related to its packing.

5.2.2 Liquid Carburizing

It uses a bath having the mixture of barium chloride, sodium cyanide, and sodium carbonate. Liquid bath causes rapid heating due to good thermal contact with components which in turn reduces the carburizing time significantly. This process overcomes both the problems of solid carburizing, as it allows better control of

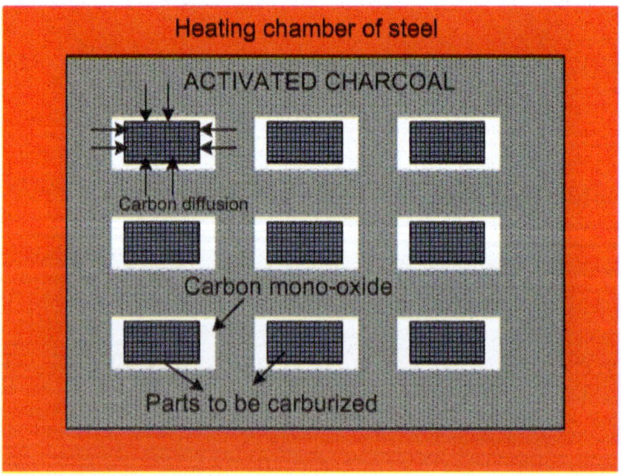

Fig. 5.5 Schematic of solid/pack carburizing process

temperature of bath. Moreover, the carburized component can be quenched directly from the bath temperature. The bath temperature and chemical reactions can be easily controlled. It takes about 30 min to 1 h to carburized layer of 500–1000 μm at 820–900 °C. Chemical reactions of liquid carburizing are given below.

$$BaCl_2 + 2NaCN \rightarrow Ba(CN)_2 + 2NaCl$$
$$Ba(CN)_2 + Fe \rightarrow Fe(C) + BaCN_2$$

Few additional reactions may also occur to introduce nitrogen in steel:

$$NaCN + O_2 \rightarrow 2NaCNO$$
$$3NaCNO \rightarrow Na_2CO_3 + NaCN + C + 2N.$$

5.2.3 *Gas Carburizing*

It uses gaseous mixture of propane and butane. These gases are allowed to flow at a high temperature (870–950 °C) around the component to be carburized, and on decomposition, these gases provide the nascent carbon atoms. The flow of gas mixture is determined by the concentration of carbon atoms. High concentration of carbon atoms in the mixture involves low flow rate. The flow rate of gaseous mixture should be such that carbon atoms available for the penetration must be equal to those penetrating the surface, otherwise shoot formation (deposition of a carbon layer on the work surface) at the surface may take place. Typical chemical reactions of gas carburizing are as under.

$$C_3H_8 2CH_4 + C$$
$$CH_4 + Fe \quad Fe(C) + 2H_2$$
$$CH_4 + CO_2 \quad 2CO + 2H_2$$
$$2CO + Fe \quad Fe(C) + CO_2$$
$$H_2 + CO_2 \quad CO + H_2O.$$

5.2.4 *Cyaniding*

Cyaniding is similar to that of liquid carburizing except that mixture of the sodium cyanide, sodium carbonate, and sodium chloride has high concentration (90%) of sodium cyanide. In this process, the surface and subsurface regions are hardened mainly due to the penetration of nitrogen (nitride formation). The penetration of carbon does not contribute appreciably in this method, whereas the reverse is true for liquid carburizing. Conversely, surface hardening by cyaniding primarily occurs due to nitride formation, whereas in case of liquid carburizing it is mainly because of martensitic transformation. Cyaniding is carried out over a range of temperature from 830 to 860 °C. Case hardened layer of 150–600 μm comprises 0.6–0.7% C and 0.8–1.2% N. Cyaniding at further higher temperatures results in larger concentration of nitrogen and carbon up to greater depths. Heat treatment is carried out after cyaniding. However, cyaniding is costlier process and uses of *the* toxic cyanide salts. Precautions are needed to protect workers and the environment from the toxic gases. The following reactions take place during the cyaniding.

$$2NaCN + O_2 \rightarrow 2NaCNO$$
$$2NaCNO + O_2 \rightarrow Na_2CO_3 + CO + N_2$$
$$2CO \rightarrow CO_2 + C.$$

5.3 Nitriding

Nitriding is based on the simple principle of increasing the concentration of nitrogen at the surface and subsurface region so as to form enough hard nitrides of iron, chromium and aluminum (Fig. 5.6). Moreover, the formation of thin layer of extreme hardness and brittle layer of nitrogen rich iron nitride (Fe_4N) at the surface and somewhat softer nitrides ($Fe_{2-3}N$) in the surface region is common. Accordingly, two zones are developed in the components subjected to nitriding: One is compound layer, and the other is diffusion layer. This process responds very effectively in the case of steels having good amount of nitride formers such as aluminum, molybdenum, chromium, and vanadium. Therefore, to increase the effectiveness of this method, a few strong and stable nitride forming alloying elements such aluminum, vanadium, titanium, and chromium are added in steels (Fig. 5.7).

Fig. 5.6 Micrograph of nitride steel samples

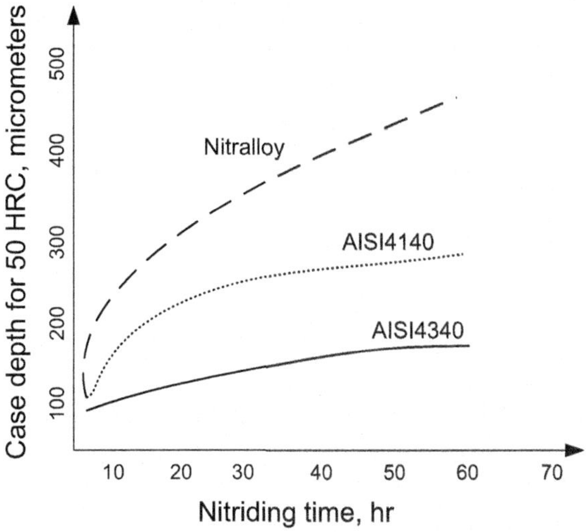

Fig. 5.7 Effect of steel composition and nitriding time on depth of case subjected to nitriding

Nitriding of steel is generally performed in the ferritic zone at about 500–550 °C primarily owing to the high enough diffusion rate for the penetration of nitrogen without having austenitic state. Nitrogen rich environment is generated by ammonia through the following reactions. The formation of thick iron nitride layer near the surface of the steel substrate (appears as white layer under microscope) is called white layer which becomes hard and brittle (Fig. 5.8). These layers get chipped off during the application which increases stress concentration and stress raisers. Therefore, white layer is generally removed first before putting the component in use. Hardening of surface reduces the crack nucleation tendency. Similarly, the development of residual compressive stress reduces the effective tensile stress which in

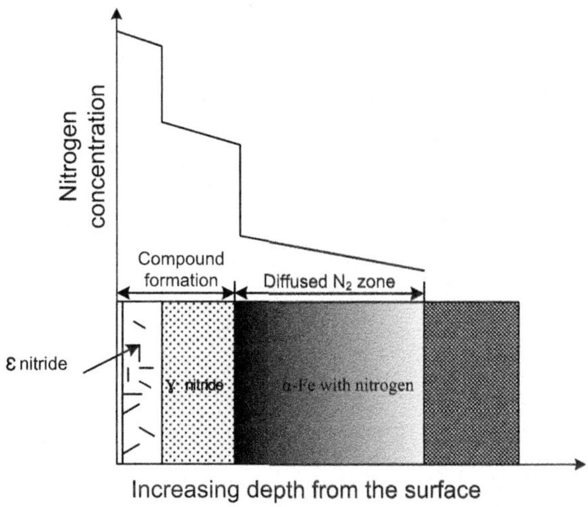

Fig. 5.8 Schematic of variation in nitrogen concentration and compound formation with increase in depth from the surface after nitriding

turn reduces the crack growth rate under fatigue load conditions. Therefore, nitriding helps to improve fatigue performance in addition to wear resistance.

$$2NH_3 \rightarrow 2N + 3H_2.$$

5.4 Plasma Carburizing and Plasma Nitriding

The basic principle of the plasma carburizing and nitriding process is similar to that of conventional carburizing and nitriding processes related to the enrichment of carbon and nitrogen respectively at the surface and subsurface layers of steel, except that in the plasma-based processes that require active carbon and nitrogen atoms are obtained directly through plasma formation under controlled conditions of temperature (300–800 °C) and pressure (300–1000 Pa) instead of passing through the number of thermal decomposition and dissociation stages of the involved hydro-carbon and nitrogen-based gases (Fig. 5.9). Therefore, plasma-based carburizing and nitriding processes are fast and offer very quick surface modification. These gases under favorable conditions of temperature, pressure, and potential difference between workpiece (cathode) and chamber (anode) result in discharge and produce plasma of gases having carbon and nitrogen in their atomic stage. The pressure of gases filled in the chamber for the production of plasma is very critical, as it affects the inter-particle spacing of plasma, which in

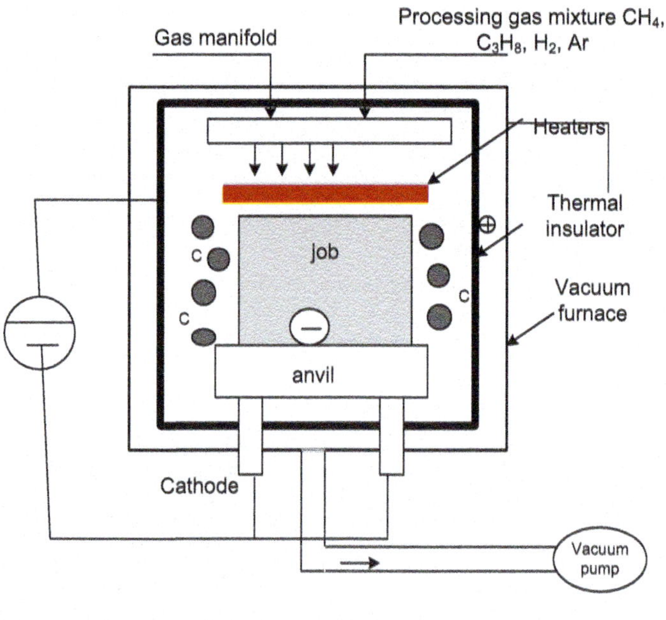

Fig. 5.9 Schematic of set up used for plasma carburizing

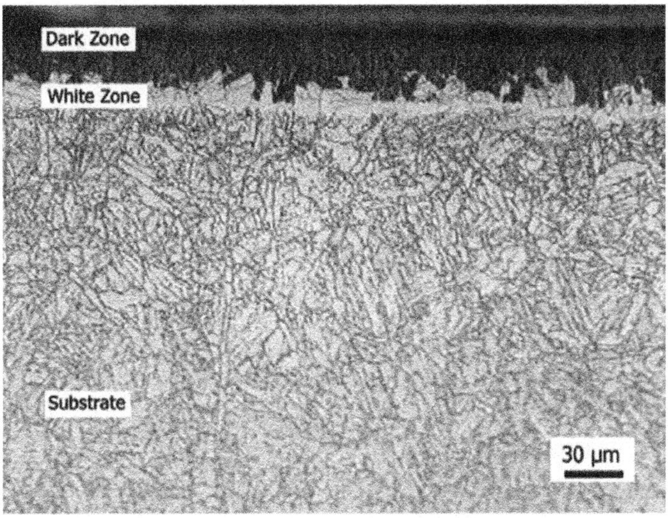

Fig. 5.10 Typical micrograph of martensitic stainless steel AISI case hardened (>1000 HV) by plasma nitride at 500 °C for 20 h (Corengia et al. 2004)

turn influences the electrical conductivity for the flow of current. In general, high pressure increases the electrical conductivity. Excessively high pressure plasma can lead to arcing, while low pressure can disrupt the discharge, which is a requisite for this group of processes to produce plasma. During plasma carburizing and nitriding process around the workpiece, a typical glowing light is produced; therefore, it is also called glow discharge plasma process. Micrograph of martensitic stainless steel AISI 410 surface case hardened up to 30 micrometers of hardness greater 1000 HV by plasma nitride at 500 °C for 20 h (Fig. 5.10). Formation of nitrides increases hardness, wear resistance and fatigue resistance appreciably.

5.5 Surface Modification by Changing Chemical Composition

The surface modification approach of surface engineering is based on the simple principle of changing the chemical composition through melting and/or diffusion. The change in chemical composition results in the formation of solid solution, intermetallic compounds, and phases which impart the desired properties. Various methods of surface modification by this approach have been grouped on the basis of the approach employed for chemical modification of near-surfaces layers.

Chemical or electro-chemical treatments involve the development of layers of phosphate or oxide at low temperature (50–100 °C). Thermo-chemical treatments (carburizing, nitriding, cyaniding, etc.) introduce carbon and nitrogen usually at a high temperature (above 500 °C). Ion implantation method changes the chemical composition at a faster rate in the presence of ions used for surface modification. Laser and electron beam alloying is based on the principle melting of substrate and alloying elements for chemical composition modification, and it is commonly used for increasing corrosion and wear resistance of surface. For example, chromium is introduced at the surface by melting, which is carried out using laser or electron beam. Pack cementation includes methods like aluminization, siliconization, chrominization that involve packing of the component to be modified in a box with suitable elements in the powder form with activators (chlorides, fluorides, etc.) for surface modification by changing the chemical composition.

5.5.1 *Laser Alloying*

The laser alloying is a simple surface modification technique of changing the chemical composition. Laser heat source is used for melting of materials to be added and substrate for the required alloying. There are two approaches of applying the materials to be added for alloying: (a) preplacement of materials to be added on the surface of the substrate and (b) feeding of stock mostly in the form of powder in the fusion zone (Fig. 5.11). The preplacement can be achieved through placement of the material that is to be added manually or by using suitable spray techniques. Manual preplacement techniques results in non-uniform thickness of preplaced materials, whereas spray process produces uniform thickness which in turn results in more uniformity in the surface chemistry modification. The concentration of alloying element in this process is dictated by dilution, that is, the extent of melting of substrate. Low dilution obtained from the shallow melting of substrate results in higher concentration of alloying element than high dilution. Hence, low dilution is preferred for surface alloying. The high energy density associated with laser or electron beam facilitates limited melting of substrate (to a very small depth); hence, it is preferred for surface alloying. However, tungsten inert gas welding can also be used for surface alloying but with high dilution (%). The laser alloying is a cost-effective method of controlled alloying of carbon steels, as the application of costly alloying steel in the form of substrate disturbs the economics. For example, Ni–P are alloyed in Al–Si alloy (A 356) by laser alloying process for improving the wear resistance. Micrograph of Ni–P alloyed at the surface of aluminum silicon alloy by laser alloying is shown in Fig. 5.12.

5.5.1.1 Laser-Assisted Nitriding

Controlled melting of the substrate using laser in nitrogen ion rich environment introduces nitrogen in the substrate which after heat treatment produced nitrides for hardening of case. Thus, the process of introducing nitrogen with help of laser is called laser-assisted nitriding (LAN). Laser-assisted nitriding of martenisitc

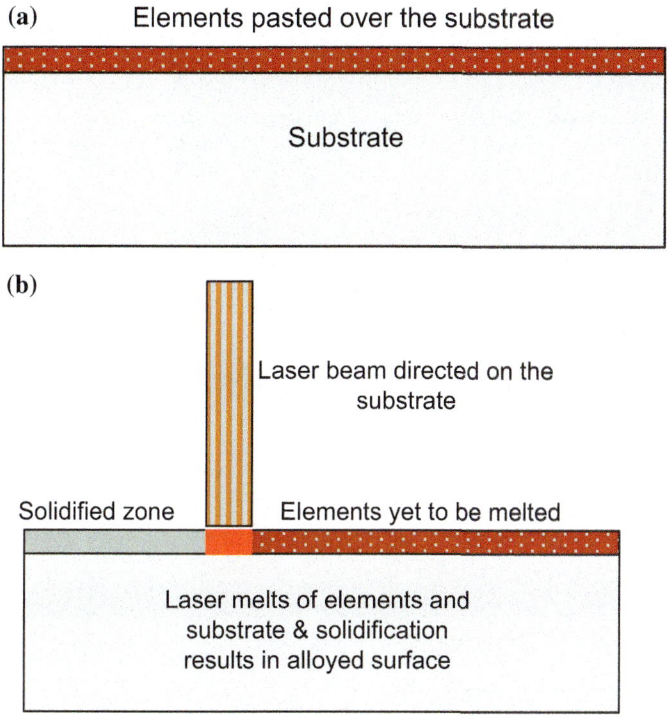

Fig. 5.11 Schematic of laser alloying **a** preplacement of alloying element/master alloys and **b** principle of laser alloying

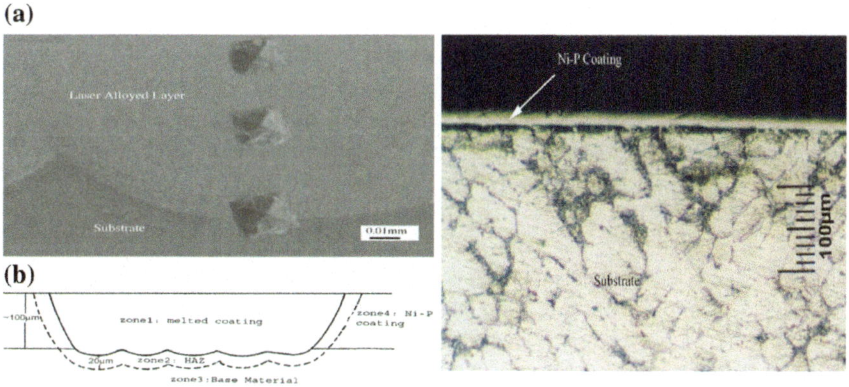

Fig. 5.12 Micrograph of Ni–P alloyed at the surface of aluminum–silicon alloy by laser alloying

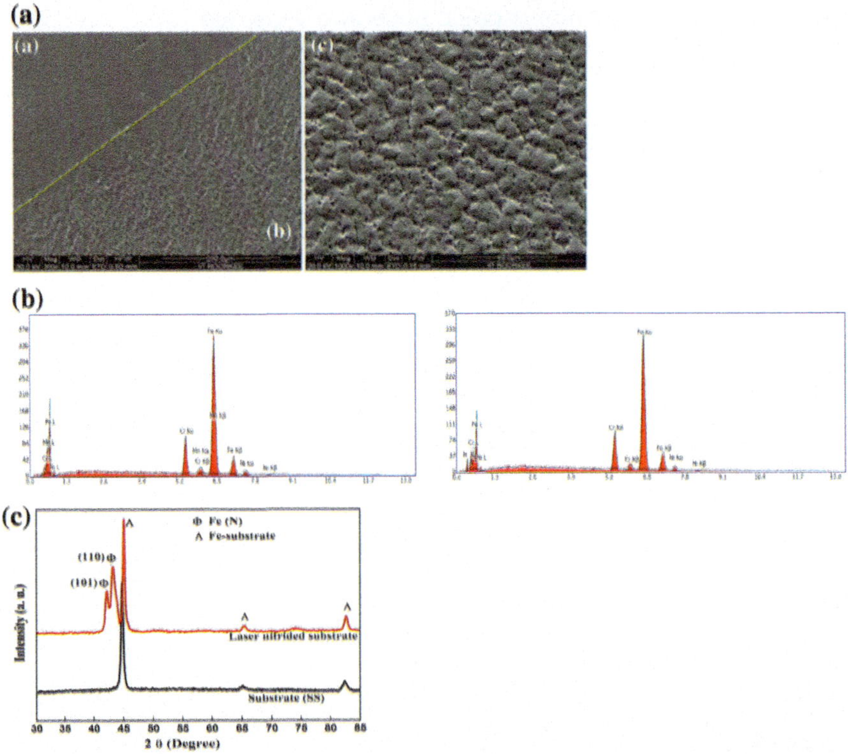

Fig. 5.13 Laser-assisted nitriding of martensitic stainless steel **a** SEM images of laser-assisted nitrided (LAN) stainless steel samples irradiated at 4 J/cm^2, **b** EDAX analysis of LAN samples, and **c** XRD analysis of LAN samples (Verma et al. 2016)

stainless steel is shown in Fig. 5.13a–c. Laser melting coupled with fast cooling results in grain refinement (Fig. 5.13a). Laser-assisted nitriding of the 13/4 martensitic stainless steel shows surface modification in terms of the structural morphology and incorporation of nitrogen in the near-surface layers (Fig. 5.13b). The EDAX by FESEM samples before and after and laser nitride indicates incorporation of nitrogen (3.4 wt%) as shown in (Fig. 5.13b). The formation iron nitrides (Fe_2N and Fe_3N) can be noticed from X-ray diffraction analysis (Fig. 5.13c). The intensity of iron nitride in the implanted samples is very low before heat treatment therefore laser nitrided samples need to be annealed for 6 h at 600 °C. LAN increases hardness by four times (from 2.74 to 8.41 GPa) as obtained from nanoindentation.

5.6 Surface Modification Using Diffusion-Based Processes (PVD, CVD)

There are a number of surface modification processes where chemical composition is changed as per the requirement by creating an artificial atmosphere having an enriched chemical compound or ions of element to be introduced into the surface of components. This is achieved through the diffusion of elements to be introduced at the surface mainly in three ways: (a) chemical reaction at the surface, (b) vapor, and (c) ions of element desired (Fig. 5.2). Accordingly, various processes are grouped under chemical vapor deposition (CVD), physical vapor deposition (PVD), and ion implantation-based processes. PVD processes work at lower temperature than CVD processes therefore CVD process causes more thermal damage of the substrate than PVD (Fig. 5.14). Further, these methods do not require metallurgical compatibility between the substrate and element to be introduced.

Material for deposition is used in the form of powders (having high surface to volume ratio), metal vapor, and plasma. Most of the above processes are performed in vacuum; hence, these are also called vacuum-based surface modification techniques. The chemical modification of very thin layer using diffusion-based techniques is mostly carried out at high temperature primarily for achieving high diffusion rate, so as to reduce the processing time (Fig. 5.15). However, high temperature promotes the

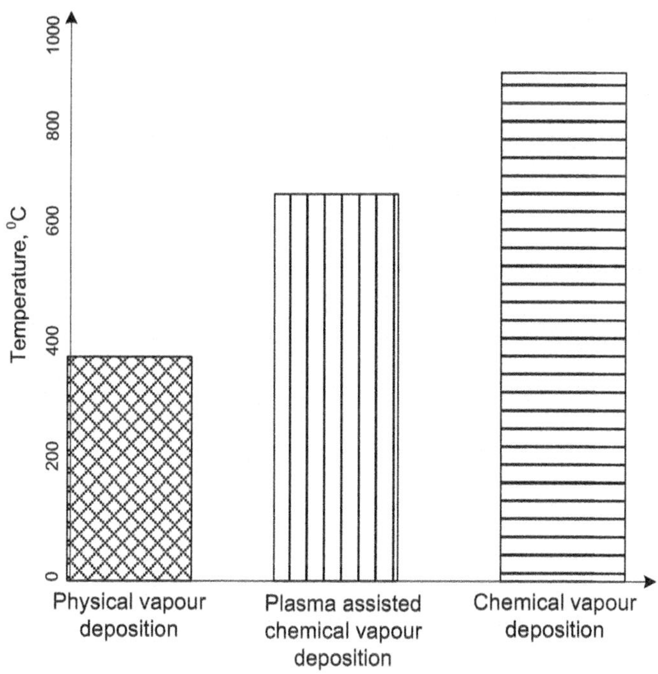

Fig. 5.14 Working temperatures of PVD and CVD processes

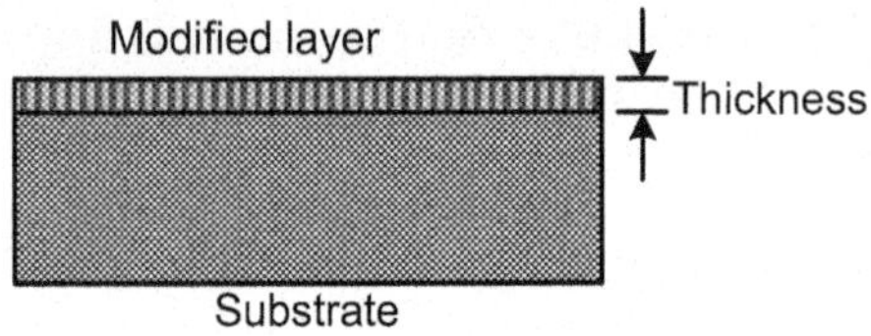

Fig. 5.15 Schematic of composition modification for surface engineering

formation of undesirable oxides and nitrides in case of the most of the metals under normal atmospheric conditions. These deteriorate the quality of modified surfaces. Hence, the processes like vapor deposition, sputtering, ion implantation, ion beam-assisted sputtering are generally performed under vacuum. Large vacuum chambers and vacuum pumps invariably form an integral part of the systems employed for the above processes. The capability of these processes to process large number of components simultaneously led to increase in the cost-effectiveness.

5.6.1 Vacuum Deposition

This is a very simple process in which the metal to be deposited is heated (using suitable electrical heating system), so that the metal evaporates. Upon evaporation, metal vapors move up at low velocity and tend to get deposited on the substrate (Fig. 5.16). At the substrate, the deposited metal diffuses gradually into the substrate to form metallurgical bond. Major advantage of this process is that it is a low-temperature process (<100 °C), hence, there is no thermal damage to substrate besides use of comparative low level of vacuum requirement. The substrate-coating bond strength is poor because of the application of low-energy metal vapor for deposition. Therefore, vapor-deposited coating (Zn, Cr, Cu on steel surface) are preferred for certain applications in which functional surfaces are subjected to low load or no load conditions such as corrosion, but it is not considered for applications where the substrate is subjected to high external loads, for example, surface fatigue, abrasion, and erosion conditions. However, vapor deposition suffers from many limitations: (a) slow process, (b) poor bonding with substrate owing to low energy of metal vapors, and (c) proper cleaning of the substrate is mandatory and no auto-cleaning of the functional surface is achieved.

5.6.2 Ion Implantation

The basic principle of the process involves bombardment of high-energy ions on the surfaces of substrate to be engineered primarily to damage the regular arrangement (lattice) of crystalline structure of the metal for achieving amorphous structure, producing well-designed critical level of dislocation density, and forming

Fig. 5.16 Vacuum deposition process **a** schematic of process principle, **b** schematic of set up, and **c** micrograph of a typical of MoS-based thin film coatings by pulsed laser deposition

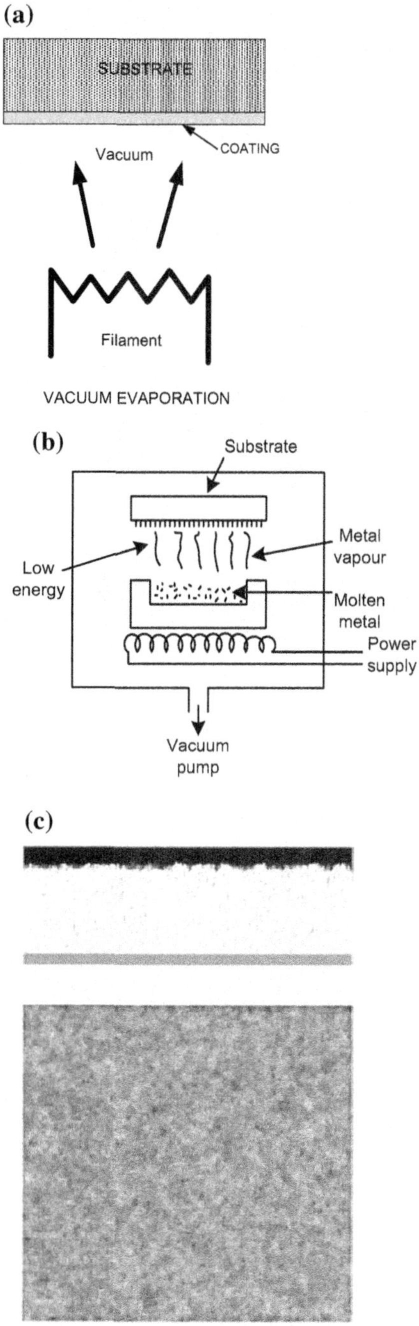

unstable and stable phases or compounds based on the interactions between implanted ions and matrix material (Fig. 5.17). These changes in near-surface layers result in increase of hardness and resistance to wear and corrosion. However, the precise mechanism of improvement in the properties of modified surface through ion implantation is not very clear. The depth of modification of structure and implantation of ions (usually 1%) is generally limited to 1 μm and the same is determined by the initial energy of ions.

Ions are produced by thermo-ionic emission mechanism using any suitable gas. Moreover, nitrogen is one of the commonly used gases for ion implantation, and accordingly, it is called nitrogen ion implantation. Ions with appropriate energy levels are filtered and accelerated through utilization of electrostatic field, following which these ions are directed on the surface so as to have about 1% concentration below surface (up to about 1 μm) by using ions 1×10^{18} to $1 \times 10^{21}/m^2$.

Inherent nature of the process, that is, use of ionic bombardment only, offers many advantages over other competitive processes: (a) low-temperature exposure of modified surfaces, (b) no dimensional and geometrical variation, (c) no adverse effect on bulk material properties, (d) selective surface modification of functional areas only can be done, and (e) self cleaning process. However, this process finds very limited applications in the industry owing to (a) the need of expertise for designing the process, (b) limited depth of modified surface (about 1 μm) which becomes deficient for many high stress applications, (c) inaccessible areas cannot be modified as it is applicable only for areas in line of sight with respect to ion beam, and (d) requirement of expensive vacuum and equipment.

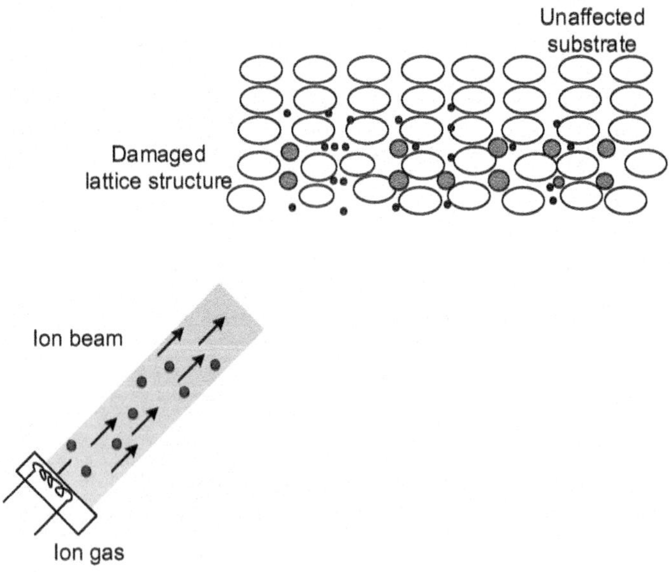

Fig. 5.17 Schematic of ion beam implantation

5.6.3 Sputtering

The sputtering is a process of release material atom by atom from the surface of target (coating material) after the impact of high-energy ions, and it is analogous to the spreading of sand particles when a ball impacts the heap of sand. Material removed from the target in the form of atoms is allowed to deposit onto the surfaces including substrate and internal chamber walls, therefore, the deposition efficiency and rate of deposition of this process is found very low (Fig. 5.18a).

A vacuum chamber filled with gas like nitrogen or argon is used for sputtering deposition. These gases are ionized and accelerated with the application of suitable potential difference (5000 V) between cathode (target coating material and substrate) and anode (a screen located between coating and substrate). The ions can be obtained from the local plasma (diode or magnetron) or a separate source of ion beam. Ion beam deposition process uses separate source ion beam (Fig. 5.18b).

Positively-charged irons strike the target coating material and substrate. The impact of ions with substrate removes the impurities and cleans the surface, whereas the impact of ions with target material releases the material atom by atom (Fig. 5.18). These atoms move toward the substrate and all around. On the way from coating to substrate, the atoms collide with argon gas and ionize them. However, these collisions of atoms removed from the target material reduce the deposition rate. Therefore, vacuum is maintained. Positively-charged screen between target and substrate accelerates the atoms and helps in deposition onto the substrate (Fig. 5.18c). High energy impact of atom on the substrate helps to strengthen the bonding between coating and substrate material. Kinetic energy of the impacting atoms of the target coating material significantly affects the characteristics of quality and deposition rate of coating. High kinetic energy imparted to atoms by imposing huge bias between screen and substrate results in high deposited rate as well as strengthens the bond between coating and substrate. However, the absence of bias results in reduction of the coating deposition rate and the bond strength between coating and substrate (Fig. 5.18d).

5.6.4 Ion Plating

Ion plating basically involves ionization of metal vapors produced by heating material which is to be coated using suitable heating system. Ionized metallic vapors are accelerated by employing potential difference between substrate and coating materials. High kinetic energy ions produced from target coating metals impact the substrate and get deposited (Fig. 5.19). However, these conditions result in high temperature during ion plating, which is considered as undesirable for substrate because it tends to impact the properties of substrate. The deposition rate

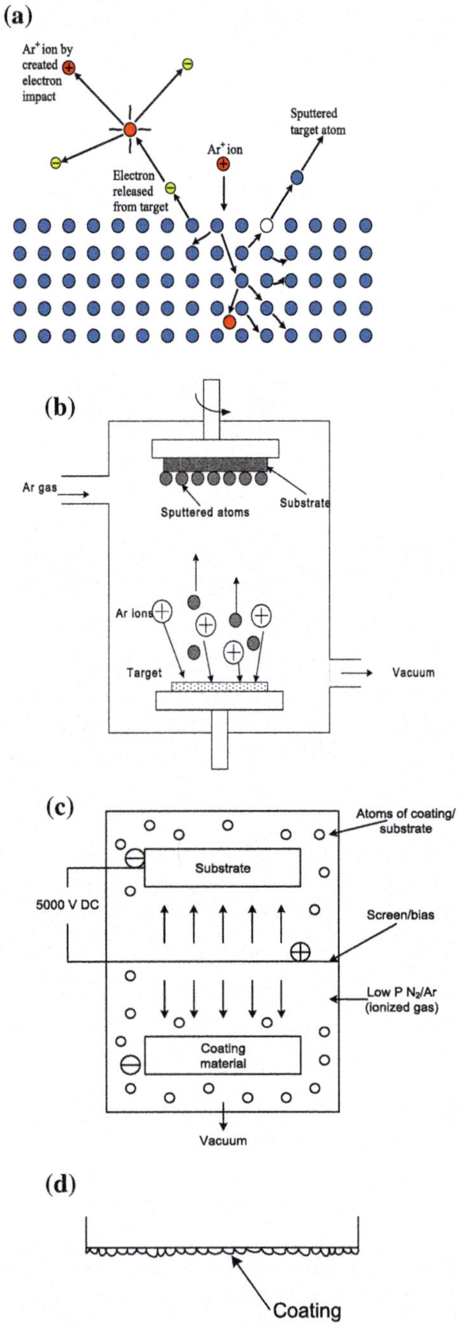

Fig. 5.18 Schematic of sputtering process **a** principle, **b** operational approach, **c** set up, and **d** coating

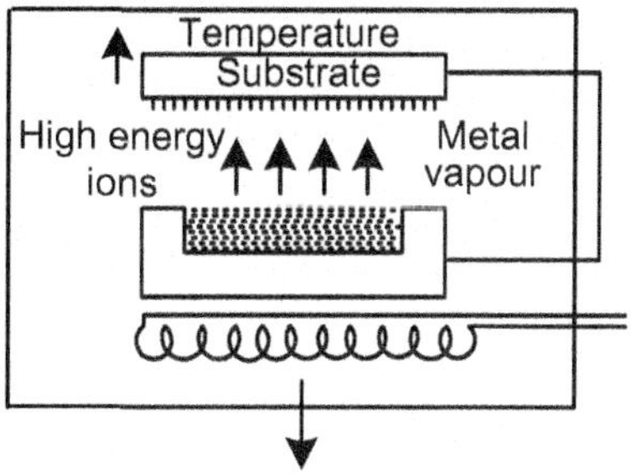

Fig. 5.19 Schematic of ion plating process

obtained from ion plating is very slow (~1 μm/h) and so it is time consuming. Further, the coating–substrate interface in ion plating is found very sharp, which can create problems associated with metallurgical incompatibility and poor bonding due to the differential thermal expansion coefficients. To deal with the problem of metallurgical compatibility, ion beam-assisted sputtering, or vapor deposition has been developed.

5.7 Ion Beam-Assisted Deposition

The ion beam-assisted deposition primarily based on the application of a separate source of ion beam along with other deposition techniques like vapor deposition and sputtering. The application of ion beam on the surface of substrate performs two roles; one is cleaning of substrate, and the other is intermixing of coating or deposited materials with substrate by using high-energy ions directed onto the surface (Fig. 5.20). Both these factors help in developing the clean, high quality, and coherent-modified surface layer. The proper mixing of coating material being deposited at the substrate overcomes the problems associated with the metallurgical incompatibility and poor bonding because sharp transition at coating–substrate interface is replaced by a continuous gradient of coating material. For example, the application of nitrogen gas ions for surface modification of steel substrate results in the formation of well mixed and coherent Fe_3N_4, which improves the surface hardness and fatigue properties.

5.7.1 *Chemical Vapor Deposition*

The chemical vapor deposition (CVD) is a group of thermo-chemical processes which performed through the usage of vapors or chemicals designed to get absorbed at the surface of substrate and is accompanied by chemical reactions at high

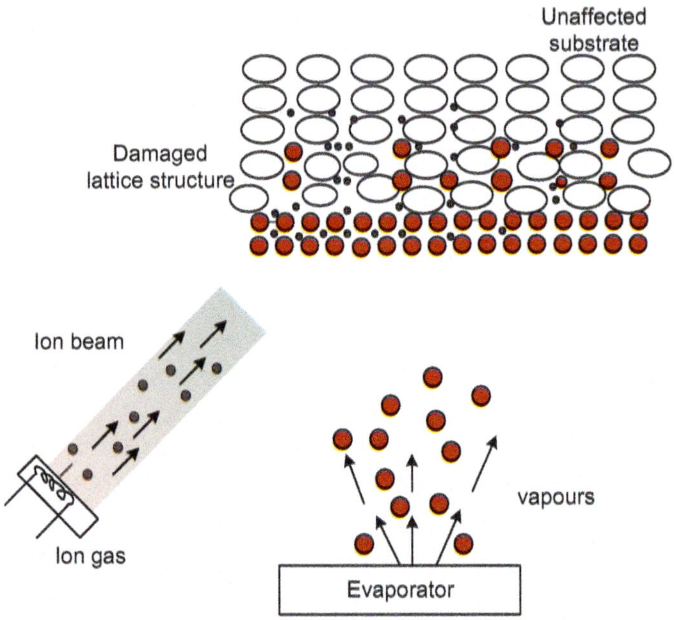

Fig. 5.20 Schematic of ion beam-assisted deposition

temperature (800–1200 °C). A chamber made of stainless steel with compartments (for accommodating the samples of the substrate to be processed) is evacuated to low pressure, and suitable gas mixture ($TiCl_4$, nitrogen, and hydrogen, etc.) is supplied for chemical reactions to take place with substrate (Fig. 5.21a). Metal cutting tool inserts are commonly developed using CVD processes. The development of titanium nitride and titanium carbide coatings on tool inserts using CVD helps in achieving high surface hardness and thermal stability. Further, CVD processes offer many advantages (a) Higher is the rate of deposition than PVD (limited to about 10 μm thickness) and (b) Increased surface hardness of ferrous materials and composite materials enhances the tensile and fatigue performance. However, the CVD processes tend to affect (and many times degrade) the structure and properties of substrate because of the requirement of high temperature exposure during CVD process. Thermal cycle used in CVD process is shown in Fig. 5.21b. Further, the advancement in CVD processes is development of plasma-assisted CVD process (Fig. 5.21c). Plasma-enhanced CVD enables required chemical reaction for CVD at much lower temperature (500 °C), thus reducing the possibility of thermal damage to the substrate. CVD coating results in significant reduction in tool wear as compared to PVD coating as shown in Fig. 5.22.

Fig. 5.21 Schematic of **a** set up for CVD, **b** thermal cycle of CVD, and **c** process principles of plasma-assisted CVD

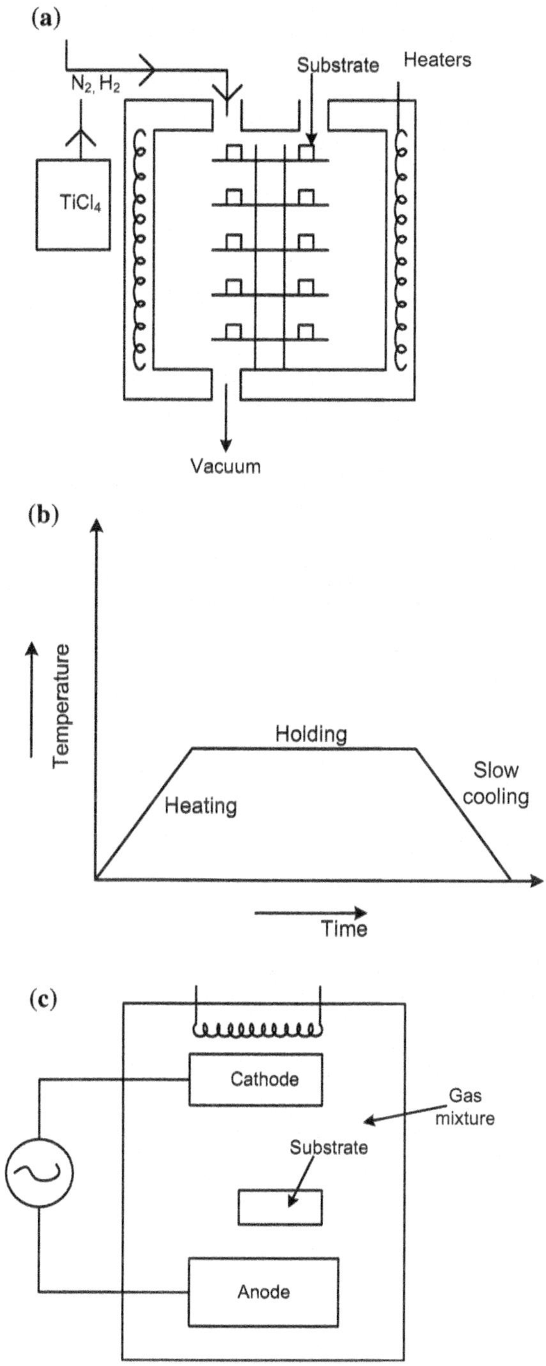

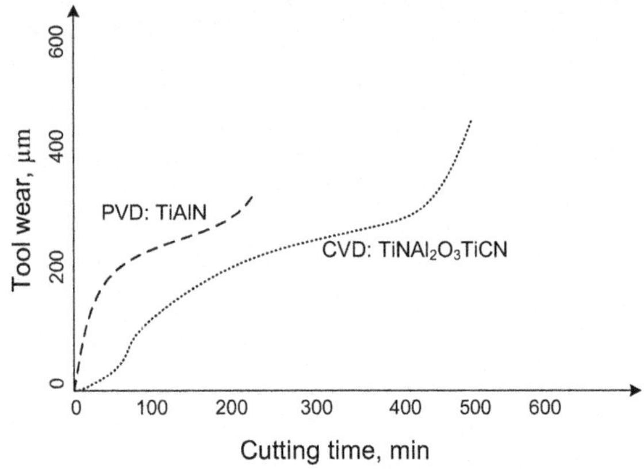

Fig. 5.22 Comparison of PVD and CVD coating on cutting tool inserts during turning of stainless steel at 290 m/min of cutting speed and 0.24 mm/rev feed rate

5.8 Boronizing

Similar to the enrichment of carbon and nitrogen for carburizing and nitriding respectively through diffusion, in case of boronizing boron is penetrated into the surface primarily for developing hard borides of different elements present in steel including of iron boride (FeB, Fe_2B). FeB is hard and brittle formed at the surface while Fe_2B is formed in the subsurface region (Fig. 5.23a, b). Boronizing increases hardness of the order of 1500–2000 VHN up to a depth of about 50 µm. The boron is introduced using boron carbide pack cementation or gas boronizing process that involves high temperature (900 °C) exposure (for about 6 h) of the components' surface to be modified in boron rich environment with the help of a pack that consisting of the mixture of boron carbide (providing boron), ammonium di-fluoride (activator), and inert matter (silicon carbide) in ratios of 5:2:93. Concentration of boron decreases gradually with increase in depth from the surface and accordingly compound formation is affected (Fig. 5.23c, d). This process increases hardness without appreciable embrittlement. However, boronizing is costlier than carburizing and nitriding for the similar kind of hardness. In general, boronizing increases hardness, abrasive and adhesive wear resistance, thermal stability, and resistance to chemical environment.

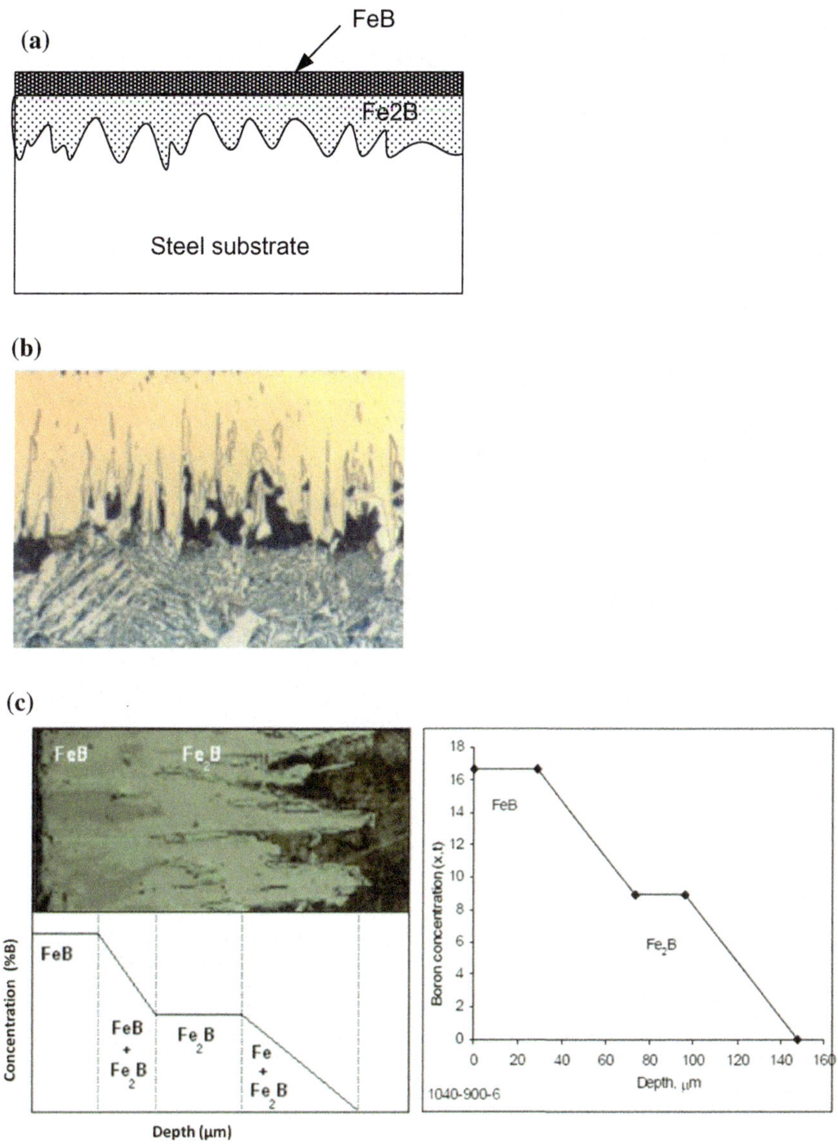

Fig. 5.23 Boronizing process **a** schematic of boride formation in the substrate, **b** micrograph of steel after boronizing showing different types of borides, **c** formation of borides and concentration of boron at varying depth from the surface, and **d** boron concentration as function of increase in depth from the surface (http://en.alper.com.tr/hizmetler/isil-islem/boronizing.html)

Questions for self-assessment

1. What are the fundamental mechanisms of the changing the chemical composition for surface modification?
2. List the common elements/species and respective methods used for surface engineering of steel by changing chemical composition.
3. Why do hardness and residual stress vary as function of depth in carburized region?
4. Explain the principle of hardening the surface of steel by carburizing.
5. Describe thermal cycle (temperature variation as function of time) used for carburizing their respective roles in case hardening.
6. Explain the following with chemical reaction involved for carburizing.

   (a) Pack carburizing
   (b) Liquid carburizing
   (c) Gas carburizing
   (d) Cyaniding

7. Explain the principle of nitriding steel. Which type of steel can be case hardened by nitriding.
8. Compare the carburizing and nitriding processes.
9. Why do carburizing and nitriding improve the fatigue life?
10. How the plasma carburizing and nitriding are beneficial over the conventional carburizing and nitriding processes?
11. Explain the principle of laser alloying using suitable schematic diagram?
12. How laser-assisted nitriding is different from conventional nitriding?
13. Compare the PVD, CVD and plasma-assisted CVD processes in respect of thermal damage to the workpiece during surface modification.
14. Explain the principle of following processes using suitable schematic diagram.

   (a) Ion implantation
   (b) Ion plating
   (c) Sputtering
   (d) Ion beam-assisted deposition
   (e) Chemical vapor deposition

15. Describe principle, method, and effect of boronizing steel on microstructure and properties.

Chapter 6
Surface Modification by Developing Coating and Cladding

6.1 Approach

The approach of surface modification by developing films, coating, and cladding is based on the simple principle of applying a good quality material on the functional surface of the substrate to achieve the desired properties such as improved corrosion and wear resistance, electrical, optical, thermal conductivity, or insulation (Fig. 6.1). The substrate is usually made of cheaper material, which lacks specific properties essential for improved functional performance and hence is subjected to surface modification by developing films, coating, and cladding of materials having the requisite properties.

These can be developed using different approaches such as chemical reaction, electrochemical reaction, complete or partial melting, mechanical interlocking, kinetic energy, and diffusion. Accordingly, different methods for developing films, coating, and cladding have been developed (Fig. 6.2). An overview of different methods has been presented in the table below.

Electrochemical treatments	Chemical treatments	Welding	Thermal spray processes	
Electroplating	Electroless plating	Oxy-acetylene weld overlays	**Combustion methods**	**Electrical methods**
Brush plating	Hot dip coating	TIG weld overlays	Flame spray	Electric wire arc spray
Anodizing	Chemical vapor deposition (CVD)	Shielded metal arc welding (SMAW)	Detonation gun (D-gun)	Air plasma spraying (transferred and non-transferred arc)
	Plasma-assisted CVD	Flux-cored arc welding (FCAW)	High-velocity oxy-fuel (HVOF)	Controlled atmosphere plasma (low pressure and inert atmosphere)

(continued)

D. K. Dwivedi, *Surface Engineering*,
https://doi.org/10.1007/978-81-322-3779-2_6

(continued)

	Laser-assisted CVD	MIG weld overlays		Liquid-generated plasma
	Fluidized Bed CVD	Submerged arc welding		
	Physical vapor deposition (PVD)	Plasma-transferred arc (PTA) weld overlays		

Fig. 6.1 Schematic of surface modification approach by formation of films, coating

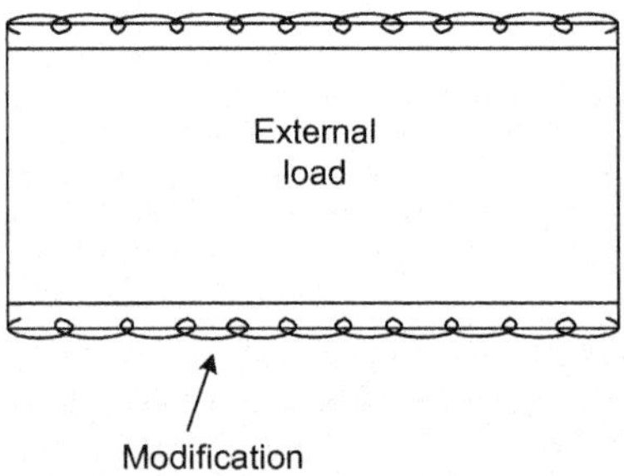

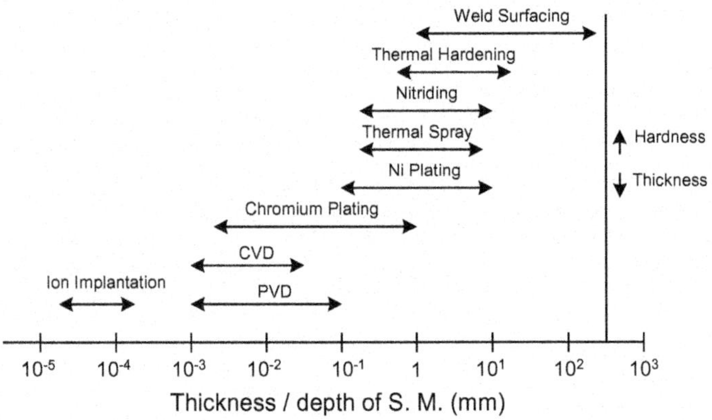

Fig. 6.2 Schematic of surface modification approach versus thickness of films, coating

The production of film, coating, and cladding all primarily involves the development of a surface layer of desired quality by utilizing the requisite material for surface modification; however, these differ with respect to thickness. Moreover, there is no clear demarcating line (in respect of thickness) to distinguish the film, coating, and cladding. In general, films are thinner (<10 μm) than coating (10–1000 μm) and cladding (>1000 μm). Increase in hardness of material reduces the thickness of layer that can be developed for better results. Harder the material, lower is the thickness of film, coating, and cladding (Fig. 6.2). Film is usually formed on the surface to achieve the thermal, electrical, and optical properties for non-stress applications; coating is generally developed for wear resistant applications, and cladding is produced for building up and reclamation purpose of worn-out component to regain the dimension and corrosion-resistant applications.

6.2 Technical Factor Affecting Performance

The performance of film, coating, and cladding for a given composition primarily depends on two aspects: (a) quality achieved meaning the extent to which they are free from impurities and undesirable discontinuities (porosity, inclusion, segregation, depletion, and interaction with atmospheric gases) and (b) extent degradation in the composition of good quality material that takes place owing to factors associated with the deposition process such as heat, intermixing with substrate metal and impurities.

The quality of buildup (as suggested by above two aspects) is largely determined by the process used for developing the same. There are two aspects dictating the ways in which sound (free from defects) buildups are made with desired combination of properties, and these aspects are related to (a) the effectiveness of the deposition process in protecting the molten metal or just deposited buildup from the atmospheric gases and (b) the effect of dilution on microstructure of the buildup owing to heat input based on the energy density available with the particulate process.

6.2.1 Protection and Dilution of Coating and Cladding

An effective protection of the molten metal pool using inert or inactive gases in processes like GTAW, GMAW, and electron beam welding helps to develop cleaner deposit than those made by using other processes such as SMAW, gas welding. Further, the intermixing of coating material with molten base metal called dilution degrades the quality of deposits. In general, the increase in dilution causes change in composition of the deposits which in turn can deteriorate their mechanical properties and so performance. Therefore, it is always preferred to have as less dilution as possible during deposition. The dilution is determined by the

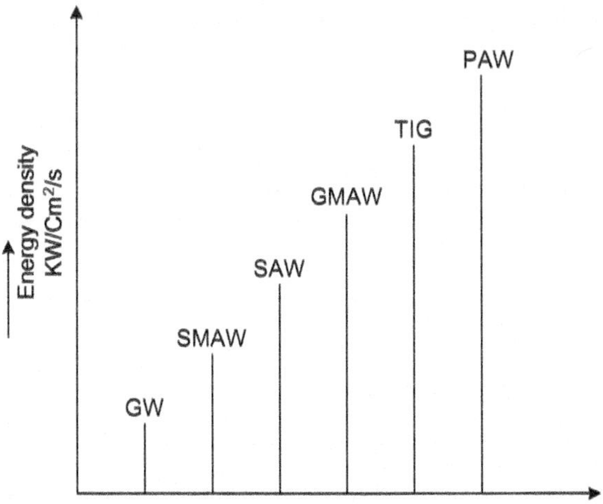

Fig. 6.3 Schematic showing relative energy density associated with different weld surfacing processes

extent (depth) to which the substrate melts during deposition. The melting of substrate to a greater depth causes more dilution and degradation in the quality of deposits. The extent of melting in general during welding surfacing of substrate and thermal spray processes is determined by the amount of heat supplied for melting of the substrate, which is governed by the energy associated with deposition (welding, spray process) process (Fig. 6.3). Higher the energy density associated with the process (like laser and plasma arc welding), lower will be the amount of heat required for melting of substrate and so lesser will be dilution.

6.2.2 *Energy Density*

Low energy associated with welding or thermal spray process increases heat input to the substrate. Large amount of heat is transferred to the underlying metal of substrate (large area), and this in turn delays the melting of substrate. However, high energy density of deposition process means energy transferred over a smaller area for a short period of time results in faster melting. Low energy density process thus would melt a larger amount of substrate leading to more dilution (Fig. 6.4).

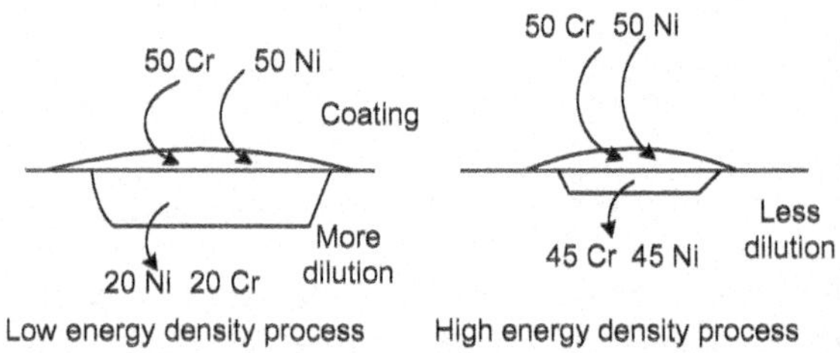

Fig. 6.4 Schematic showing the effect of energy density on dilution of weld surfacing

6.2.3 Cooling Rate

Apart from dilution, the cooling rate experienced by coating and cladding during solidification is another important technical aspect affecting the grain structure as well as their mechanical and tribological performance. It is well established that high cooling rate produces fine grain structure and improved mechanical properties. Therefore, all the factors associated with a particular coating and cladding process such as heat input, initial plate temperature, thermal conductivity, and other physical properties of materials lowering of the cooling rate, will eventually coarsen the grain structure of the coating and cladding, while cracking tendency and embrittlement of substrate and coating/cladding is reduced due to low cooling rate in case of high hardenability steel.

The conventional welding processes like GTAW and GMAW with pulse current option reduce the heat input through pulsing the current between peak and base current level. Melting facilitated during the supply of pulse current and solidification occurs at the base current level. Interrupted melting and cooling reduces the size pool and dilution, but increases the cooling rate owing to reduce heat input. Therefore, pulse GMAW and pulse GTAW offer cladding and coating of better performance than those produced using conventional processes (without pulsing).

6.3 Metals for Developing, Coating, and Cladding

Depending upon the approach being used for developing film, coating, and cladding, materials can be in the form of chemical solution, gaseous, wire, strips, electrodes, and powders. Different material systems and their alloys for specific applications have been described in Chap. 3. The forms of material used in different processes for developing buildups are shown in the table below.

S. No.	Process	Form of material
1	Electroless plating	Solution
2	Thermal spray process (flame spray, plasma arc spray, detonation spray, arc spray)	Powder, wire
3	Welding process (SMAW, SAW, GMAW, GTAW)	Electrode, wire, strips
4	Electron and laser	Powder
5	Hot dip method	Molten bath

The electrolysis based methods such as electroplating can be applied to substrate having electrical conductivity, while electroless method can be used for developing coating of nickel and nickel-based composites onto both the electrical conducting and non-conducting substrates. Welding-based methods are applied for developing coating and cladding onto the substrates of thickness usually greater than 5 mm and with reasonably good thermal and electrical conductivity. The entire range of metals systems of commercial importance like wrought iron, steel, cast irons, aluminum, magnesium, copper, brass and bronzes, titanium, and nickel-based alloys can be subjected to surface modification by developing coating and cladding for improved performance during the service.

6.4 Weld Surfacing

The surface modification by welding of a surface of substrate basically involves deposition of weld bead on the surface of substrate commonly known as the bead-on-plate weld. The bead-on-plate weld is deposited primarily by melting the material to be deposited (in the form of filler wire or electrode) on the near surface layers of substrate using suitable heat source. Attempts are made to melt the substrate as less as possible, but it should be deep enough so as to develop metallurgical bond between the substrate and bead-on-plate weld. Accordingly to the process of welding, heat may be generated in different ways: (a) combustion of oxy-fuel gas mixture in the case of gas welding, (b) establishing a welding arc between electrode/filler (being used as coating or cladding material) and substrate, for example SMAW, GMAW, SAW, PAW, and (c) external source of heat like laser or electron beam. The energy density associated with the particular heat source determines the amount of heat required for melting of substrate that will be used during welding.

For example, gas welding is a low energy density process; therefore, it supplies large amount of heat to the substrate which in turn causes excessive melting of the substrate resulting in high dilution of coating and cladding as well as low cooling rate during solidification of the bead-on-plate weld deposited. These factors degrade the quality of bead-on-plate weld deposited for surface modification due to (a) poor control over the composition due to excessive dilution and coarse the microstructure due to slow cooling rate caused by high heat input. Therefore, efforts are made to reduce the heat input for weld surfacing using high energy density welding process (if possible) and selection process parameters accordingly.

6.4.1 Gas Welding

Gas welding is comparatively high heat input process and uses the heat generated by combustion of the fuel gas and oxygen. The oxy-fuel gas combustion flame is composed either of two or three cones which are spread over large surface area of the substrate; therefore, energy density associated with gas welding is very low, and hence, it transfers lot of heat to the underlying metal of the substrate and causes greater dilution and slow cooling of the bead-on-plate weld. Therefore, the

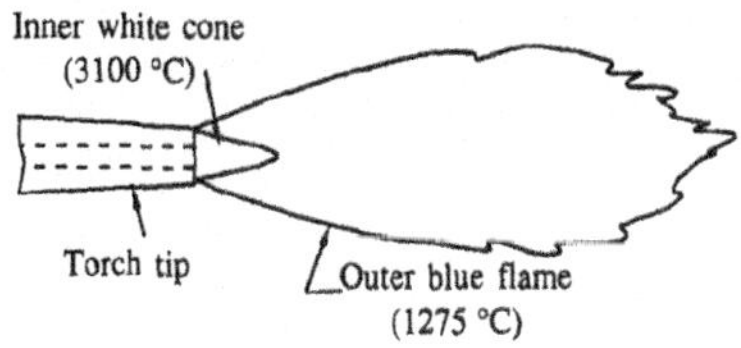

Fig. 6.5 Schematic of neutral flame in gas weld surfacing showing temperature in different zones

application of gas welding for weld surfacing of thick and heavy section plates and components using large diameter filler becomes difficult. Moreover, gas welding is utilized for reducing the cooling rate experienced by the weld pool and heat-affected zone during weld surfacing of hardenable steels and this helps in avoiding the formation of hardness and brittle structure and in lowering the cracking tendency. There are three types of flames formed during gas welding depending on fuel and oxygen ratio: (a) oxidizing, (b) carburizing, and (c) neutral flame (Fig. 6.5). The application of carburizing flame for weld surfacing causes enrichment of the weld bead with carbon, which in turn increases the hardness and even sometimes leads to embrittlement of the weld bead. This process is used for developing welding surface of nickel and nickel–tungsten carbide composite, beside many common iron-based surfacing metals.

6.4.2 Shielded Metal Arc Welding

Shielded metal arc welding (SMAW) is one of the most commonly used weld surfacing processes owing to the ease of availability, flexibility, simplicity, and cost effectiveness. It uses electrodes composed of core wire and flux coating. Core wire provides metal for developing weld surfacing and protection of the solidifying weld pool that comes from the thermal decomposition of flux coating producing inactive gases in the form of carbon dioxide. Flux is also modified using alloying elements and hard constituents such as chromium carbide, tungsten carbides for developing weld bead of desired composition and characteristics.

The dilution during the weld surfacing by using SMAW can be 5–10% depending upon the welding parameters, namely welding current, arc voltage, and welding speed. An increase in welding current and reduction in welding speed enhances the heat input unit length, which leads to increase in the dilution. Hence, the selection of welding parameters is preferred such that dilution in minimized while offering reasonably good deposition rate and enough penetration of the substrate for developing metallurgical bonding between bead and substrate (Fig. 6.6). However, the quality of the weld surfacing deposited using SMAW process is found somewhat poorer than what is obtained by other arc welding process, namely GMAW, FCAW PAW, etc., due to better protection to the weld pool from atmospheric gases. Further, the deposition rate offered by SMAW does not favor the application of this process by large-scale surfacing of wide surface of

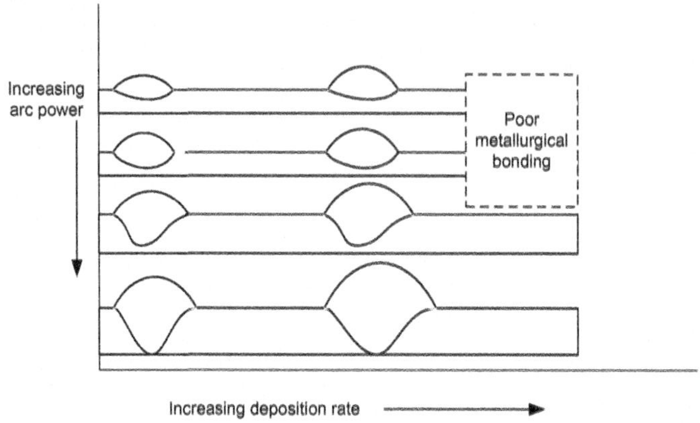

Fig. 6.6 Schematic of arc power and welding speed on bead geometry

huge industrial systems because of the intermittent nature of the process, as it uses the electrode of limited length 300–450 mm, which is consumed during weld surfacing and needs replacement regularly.

This process is extensively used for cladding, hardfacing, and buttering the surfaces for different purposes because ease of use, flexibility to apply metal at desired places, simple equipment. Cladding involve developing a thick layer of metal being applied primarily for corrosion resistance applications. Hardfacing is used for developing a layer of hard material for improving the wear resistance under abrasive, adhesive and erosive wear conditions. A butter layer is applied to isolate the base metal from the weld metal for dealing with problems related with metallurgical incompatibility and large difference in properties of base metals being joined.

Heat input and number of layers determine the dilution of each layer which in turn affects the chemical composition, hardness and wear resistance of surfacing (Fig. 6.7). Increasing heat input by increasing welding current or decreasing the welding speed increases the dilution. Increase in dilution degrades the chemical composition and properties of the weld surfacing. Weld surfacing of mild steel using high Cr high C electrode by shielded metal arc welding shows that increase in

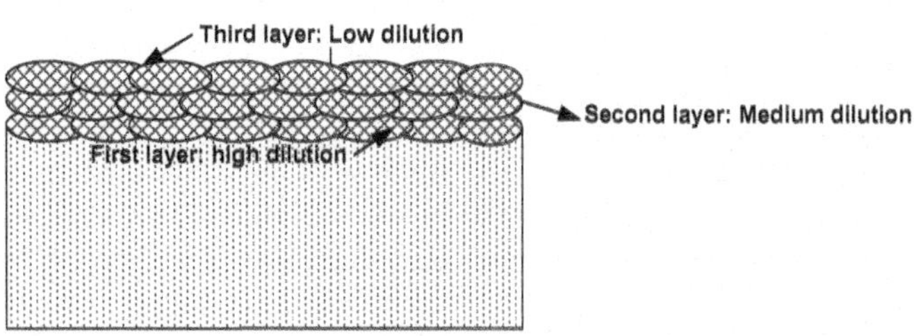

Fig. 6.7 Schematic of number of passes and dilution in weld surfacing

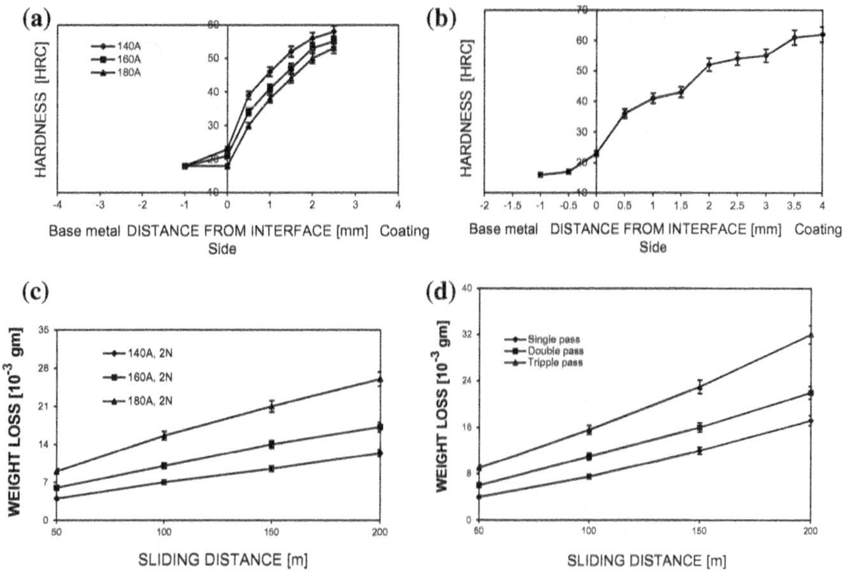

Fig. 6.8 Effect of heat input (welding current) and number of layers on hardness and abrasive wear resistance of Fe–Cr–C weld surfacing **a** hardness for different welding current, **b** hardness for different number of passes, **c** abrasive wear for different welding current, **d** abrasive wear for different number of passes (Dwivedi 2004)

heat input by raising the welding current reduces the hardness and wear resistance, while increase in number of passes does reverse (Fig. 6.8a–d). Reduction in hardness and increase in wear of weld surfacing of Fe–Cr–C system is attributed to decrease in concentration of alloying element, e.g., Cr and C, in the weld surfacing which in turn affect metallurgical structure and so hardness and wear resistance in two ways (a) reduction in hardenability, i.e., ease of martensitic transformation in iron-based wear beads and (b) reduction in fraction of hard microconstituents, namely eutectic carbide (Fe–$M_{23}C_6$) and primary carbide ($M_{23}C_6$). Eutectic carbides offer good wear resistance against adhesive wear and fine abrasive and erosive particles while coarse primary carbide particles result in increased hardness and improved wear resistance against coarse abrasive particles.

6.4.3 Submerged Arc Welding

Submerged arc welding (SAW) process uses heat generated by an electric arc between bare copper-coated consumable electrode wire and the substrate. In this process, welding arc and the weld pool are completely submerged under the cover of granular and fusible flux; therefore, it is called so. During welding, granular flux is melted by the heat generated through arc and it provides protection to the weld pool contamination from the atmospheric gases. The molten flux reacts with the impurities in the molten metal to form slag and offers the following effects on the weld joints.

- Increased cleanliness of weld metal and so improved properties of weld surfacing.
- Molten flux becomes lighter than molten metal; hence, it floats on the top of solidifying metal; therefore, it is required to protect the molten pool contamination from atmospheric gases.
- Shielding of the molten pool by molten flux and solidified slag and un-melted flux retards the cooling rate of the molten pool and HAZ which in turns decreases the cracking tendency of hardenable steel.

The SAW process is known to be of high heat input and high deposition rate process; therefore, it results in high dilution level of weld surfacing and causes more degradation in the quality of weld beads (Fig. 6.9). High deposition rate makes it suitable for weld surfacing of large surface area.

Bead geometry and depth of penetration are the two important characteristics of the bead on plate weld that are influenced by size of the electrode for a given welding current setting. In general, an increase in size of the electrode decreases the depth of penetration and increases width of weld bead for a given welding current (Fig. 6.10).

Submerged arc welding is used for weld surfacing of large surface areas wherein flux containing alloying element is fed continuously through hopper and is melted

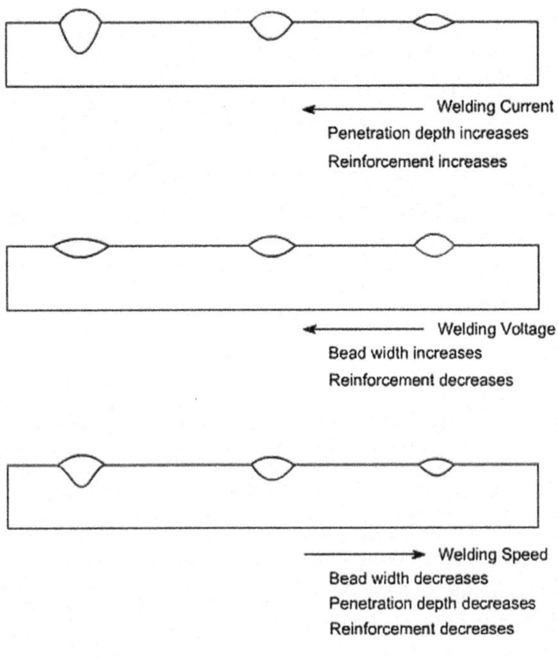

Fig. 6.9 Schematic of effect of arc voltage, current, and welding speed on bead geometry

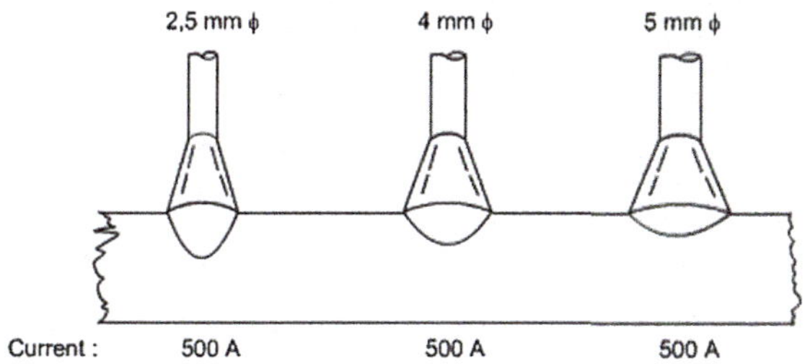

Fig. 6.10 Schematic of the effect of electrode diameter on bead geometry

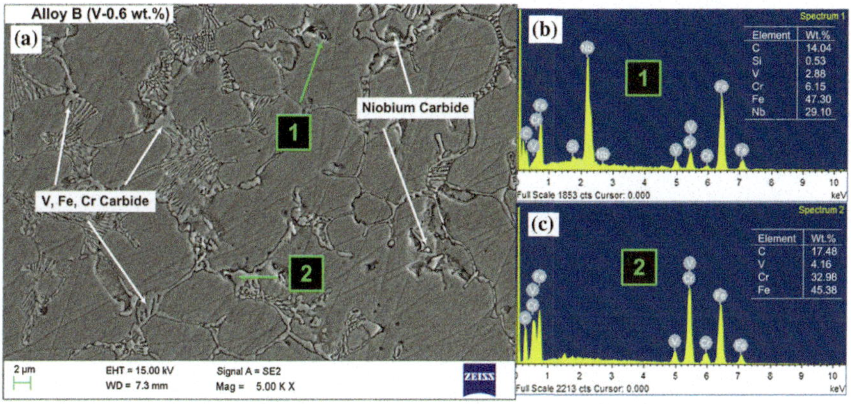

Fig. 6.11 Typical hypoeutectic Fe–Cr–V–C system **a** SEM micrograph showing various carbides and **b** EDAX analysis of weld surfacing (Singla 2016)

using heat of the arc established between continuously fed consumable electrode (usually low carbon steel) and substrate for weld surfacing of structural steels. Fluxes for hardfacing usually comprises Fe–Cr–C, Fe–V, Fe–Mo, Fe–Nb, boron, graphite, and sodium/potassium silicates. However, many technical aspects such as segregation tendency of few elements in powder mixture being fed, depth of penetration (and so dilution), poor protection of metal weld bead from atmospheric gases leading to increased loss of the alloying element from the weld bead during surfacing, etc., influence the consistence of composition, hardness and wear resistance of SAW-deposited weld surfacing. Therefore, a very close control over the composition of alloying elements in flux, welding parameters, and dilution from melting of substrate and filler is needed to realize the desired composition and wear resistance of such weld surfacing. Due to dilution, the microstructure of SAW weld surfacing of Fe–Cr–C system can be hypereutectic at the top, eutectic in the middle, and hypoeutectic close to the weld interface (Fig. 6.11).

Borle et al. (2015) investigated that effect of AC balance (50–75%) on dilution and microstructure of Fe–Cr–C weld surfacing deposited on steel substrate. AC balance

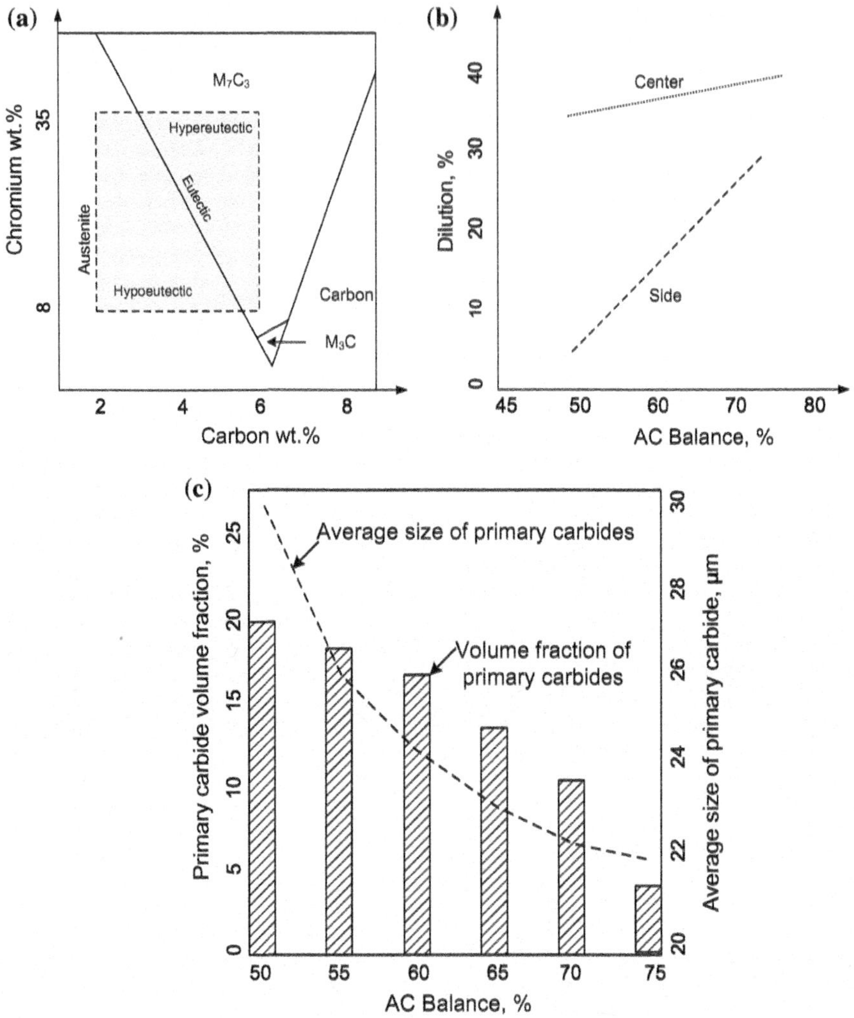

Fig. 6.12 **a** Schematic of relationship between Cr and C content on phase structure in Fe–Cr–C system, **b** effect of AC balance on dilution level and **c** effect AC balance on fraction and average size of primary carbide particle hypereutectic F–C-System (Borle et al. 2015)

indicates fraction of time electrode is positive (positive polarity) in a cycle. Increase of AC balance increases heat input from 3.82 to 4.30 kJ/mm and dilution along the centerline increase from 3.7 to 31.1% which in turn decreases the volume fraction of primary carbides from 21 to 3% (Fig. 6.12a–c). Decrease in volume of fraction with increase in AC balance (%) is attributed to increase of dilution of weld surfacing. The increase in dilution with AC balance is due to the metal spending increasing amounts of time as a non-thermionic cathode, which is associated with more heat input on the surface. Further, average diameter of carbides decreased from 30.3 to 21.8 μm with the increase in AC balance (Fig. 6.12c). The increase in volume fraction of carbides with decrease in balance therefore should increase the wear resistance. The use of AC waveform with appropriate balance (near 50%) results in a microstructure which is expected to provide the best wear resistance.

6.4.4 Gas Metal Arc Welding

Gas metal arc welding (GMAW) is based on the principle of developing bead-on-plate weld by melting of the material to be applied and faying surfaces of the substrate using the heat produced by a welding arc established between the substrate and a consumable electrode. The quality of the weld surfacing developed using GMAW is found better than those developed using gas welding and SMAW because arc and weld pool are well protected by a jet of shielding inert or inactive gas, which comes out of the nozzle and forms a shroud around the arc and molten pool. Metal inert gas process uses the automatically fed consumable electrode that is to be used for weld surfacing purpose; therefore, it offers high deposition rate and is suitable for good quality weld surfacing by GMAW required for industrial fabrication (Fig. 6.13). Consumable electrode is fed automatically, while torch can be controlled either manually or automatically. Therefore, this process is found more suitable for weld surfacing of comparatively thicker plates of reactive substrates (Al, Mg, stainless steel) and ferrous metals because otherwise the quality of weld bead is adversely affected by atmospheric gases at high temperature. Further, energy density of GMAW is higher than SMAW process; thus, dilution and adverse effects of weld thermal cycles on weld surfacing and underlying substrate can be controlled in a better way by reducing the heat input.

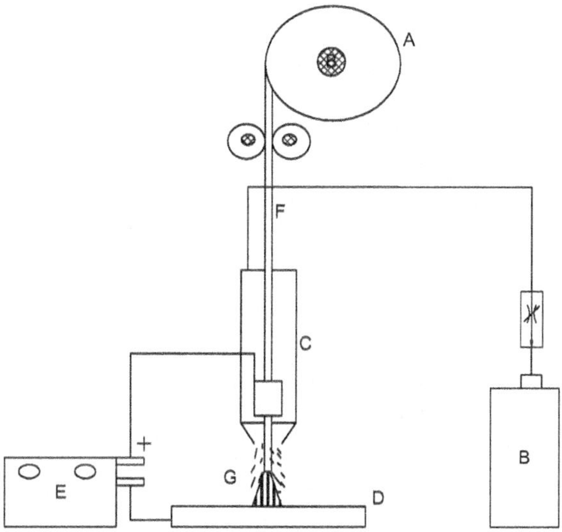

Fig. 6.13 Schematic of GMAW process showing important elements: (A) welding spool, (B) shielding gas cylinder, (C) welding torch, (D) base plate, (E) welding power source, and (F) consumable electrode

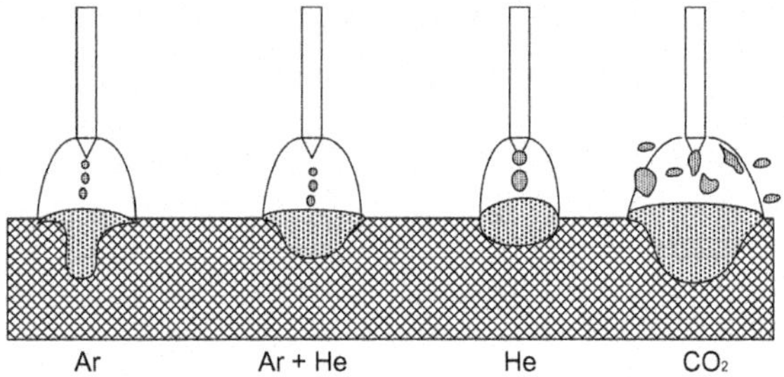

Fig. 6.14 Schematic showing the influence of shielding gas on mode of metal transfer

Shielding gases such as Ar, He, CO_2 and their mixtures are used for protecting the molten metal from the atmospheric gases. Shielding gases in GMAW also affect the mode of metal transfer (Fig. 6.14). GMAW with Ar as shielding gas results in significant change in the mode of metal transfer from globular to spray and rotary transfer (with minimum spatter) with change of welding current, whereas He mainly produces globular mode of metal transfer. GMAW with CO_2 results in weld bead with a lot of spattering. Shielding gas also affects the width of weld bead and the depth of penetration owing to their influence on heat generation during weld surfacing.

Among various weld surfacing parameters such as welding current, voltage, and speed; probably welding current is the most influential parameters affecting penetration and so dilution, deposition rate, weld bead geometry, and quality of weld bead. However, arc voltage directly affects the width of weld bead. An increase in arc voltage in general increases the width of the weld. Welding current is primarily used to regulate the overall size (cross-sectional area) of weld bead and dilution. Too low welding current results in the pilling of molten metal on the faying surface as weld bead instead of penetrating the workpiece. These conditions increase the reinforcement of weld bead without enough penetration. Excessive heating of the workpiece because of too high welding current may cause distortion due to residual stresses. Optimum current gives optimum penetration/dilution and weld bead width.

6.4.4.1 Pulse Gas Metal Arc Welding

Pulse GMAW is a variant of metal inert gas welding. It is based on the principle of pulsation of welding current between a high and a low level at regular time intervals (Fig. 6.15). High-level welding current is termed as peak current and is primarily used for the melting of electrode and faying surfaces of the substrate, whereas low current is generally called background current which performs two functions; one is the maintenance of the welding arc, and the other is providing time for solidification of the molten pool through dissipation of the heat to base metal. This feature of current pulsation reduces net heat input to the base metal during weld surfacing which in turn facilitates the bead-on-plate weld with very low dilution, especially for developing beads on thin sheets and odd positions.

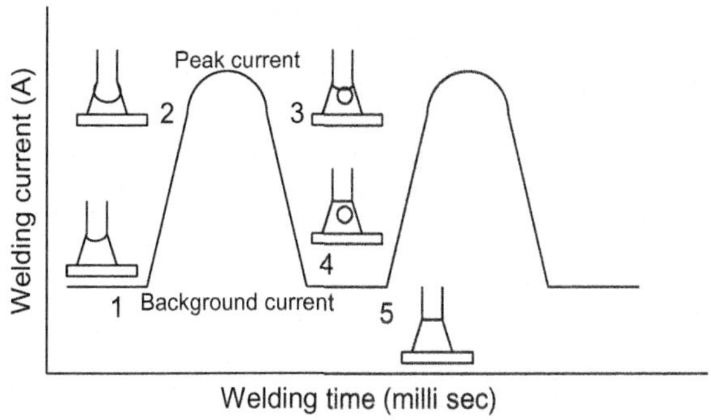

Fig. 6.15 Schematic showing the relationship between the welding current and time with metal drop formation tendency in pulse GMAW

Murugan and Parmar (MPT 1994) emphasized that a successful surfacing by GMAW requires optimization of process parameters to have low dilution and crack-free weld bead. The work on surfacing of structural steel by austenitic stainless steel (AISI 316L) using submerged arc welding revealed that increase in nozzle to workpiece distance significantly decreases the dilution, while increase in open circuit voltage increases width of weld bead and marginally affects the penetration. In line with hot wire gas tungsten arc welding, recently, an attempt was made to develop weld surfacing by preheating the electrode using auxiliary source in GMAW so as to decrease heat input (low welding current) for lowering the dilution.

Vespa et al. (2012) used controlled short-circuit gas metal arc welding process for weld surfacing of Ni–WC surfacing on the structural steel substrate using very low heat input ranging from 10 to 110 J/mm so as to reduce the thermal degradation of tungsten carbide particles. Short-circuit-controlled GMAW helps to avoid undesirable decarburization of the WC particles during weld surfacing which is commonly observed in thermal spray coating. A reaction layer of WC is formed around the reinforcing phase. An increase in heat input increases average thickness reaction layer (from 3.8 to 7.2 μm) and dilution of the reinforcing phase, while decrease in hardness of the deposited coatings from 587 to 410 HV_{10} was observed (Fig. 6.16a). Backscatter electron image showing the original WC/W2C particles with reaction layer (marked A) and the morphology of the newly formed WC particles (marked B) are shown in Fig. 6.16b.

6.4.5 *Flux-Cored Arc Welding*

Weld surfacing using flux-cored arc welding (FCAW) process is similar to that of GMAW process except that FCAW provides protection to the weld bead through thermal decomposition of flux present in core of the electrode, whereas in case of GMAW process as inert or inactive gas shroud is developed around the molten pool for protecting the pool from atmospheric gases. FCAW electrode is sometimes filled with desired microconstituents required in the weld bead for surfacing

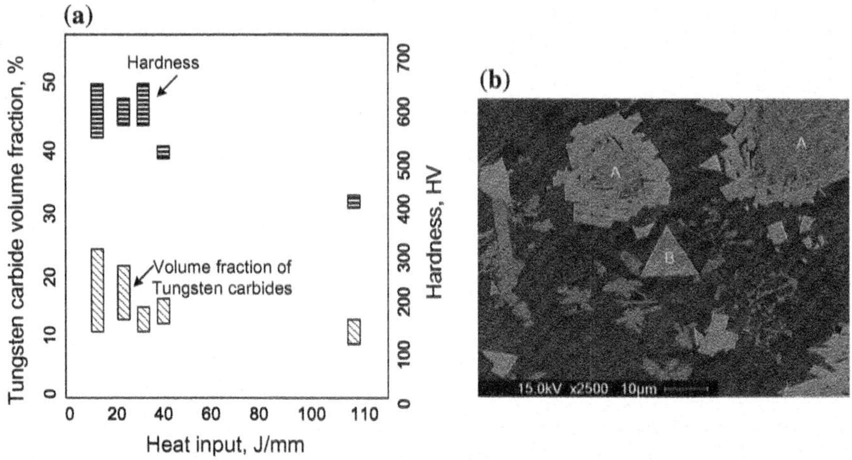

Fig. 6.16 For WC/Ni-based coatings deposited by controlled short-circuit MIG welding **a** effect of heat input on hardness and fracture of tungsten carbide particles and **b** BSE image of coating (Vespa et al. 2012)

purpose. This process has also been referred as metal-cored arc welding because desired constituents, namely metallic alloying elements and other components (such as 1–2% basic fluxes needed for generated inactive shielding gases and facilitate arc initiation and maintenance) are kept in core of tubular electrode. During weld surfacing, heat of arc established between tubular electrode and substrate is used for melting of the substrate and alloying elements within the tubular electrode.

During the last 20–30 years, there has been a worldwide trend toward replacing SMAW with other processes that allow a higher deposition rate and lend themselves better for automation. Continuous fed wires are being increasingly used. Among such processes, flux and metal-cored wires can be produced in relatively small quantities of electrodes with different chemical compositions. Among the different cored wire types, those used under gas protection can be classified as flux-cored and metal-cored wires. Metal-cored wires provide improved penetration, smooth arc, low spatter level, higher deposition rates, and the most important easier to use than solid wires. Metal-cored wires present an important technological advantage due to the near absence of slag, which makes them adequate for robotic welding (Myers 2002; Lyttle 1996).

The metal-cored wire has a composite structure made of outer metallic sheath of a cored wire and inner part (core) composed of alloy powder having required elements like Mn, Si, Ni, Cr, Mo, Fe, and arc stabilizers such as Na, K (Fig. 6.17). The outer metallic sheath of cored wire conducts current for welding. Composite nature of cored wire provides higher current density (compared with solid wires) which results in greater deposition rates even at equal current level. Internal component/contituents of cored wires can be formulated for specific applications

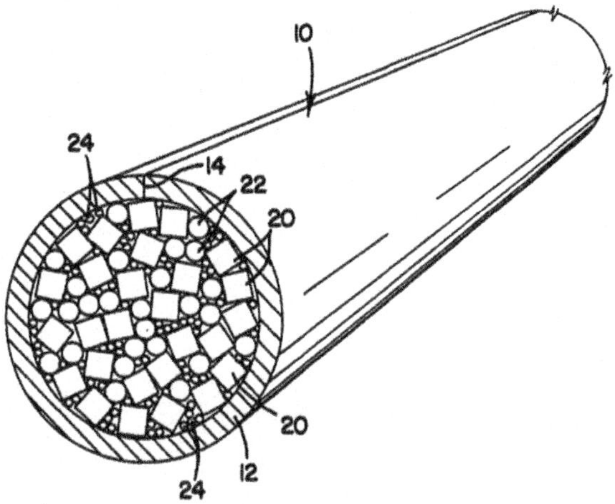

Fig. 6.17 Schematic diagram showing metal-cored wire for GMAW welding (Chai 1993)

with the metal-cored wires. These have little or no slag forming ingredient in the internal fill of wire. This permits multipass welding without need of deslaging.

6.4.6 Gas Tungsten Arc Welding

The gas tungsten arc welding (GTAW) process uses the heat generated by an electric arc between substrate and non-consumable tungsten electrode for the melting of substrate and filler material to be applied for weld surfacing. Many favorable features are associated with GTAW which include small arc gap, low welding current, and intense arc heat (owing to high energy density), and effective protection of weld pool by inert gases makes it suitable for developing quality weld surfacing with low dilution (<5%) and cleaner weld beads. Low heat input lowers the dilution and deposition rate and even sometimes making it difficult to melt large diameter filler wires for weld surfacing. To overcome the problem of low deposition (while exploiting the benefit of low heat input in the form of low dilution), hot wire GTAW has been developed, as in this process, the filler wire is preheated to lower the heat requirement for melting before being fed into the arc region. Like the GMAW process, pulse variant of GTAW process can be effectively used to reduce the net heat input for the development of weld surfacing, which in turn refines the grain structure and lowers the dilution and decreases adverse effects related to weld thermal cycle such as residual stress and distortion.

Cladding of Colmonoy 6 on the steel substrate developed using tungsten inert gas welding process improves the sliding wear resistance. Typical SEM macrograph of Colmonoy 6 shows sound thick coating (Fig. 6.18a–c) and micrograph revealed defect-free metallurgical bond at the cladding-substrate interface (Fig. 6.18b) and dendritic structure away from the interface (Fig. 6.18c).

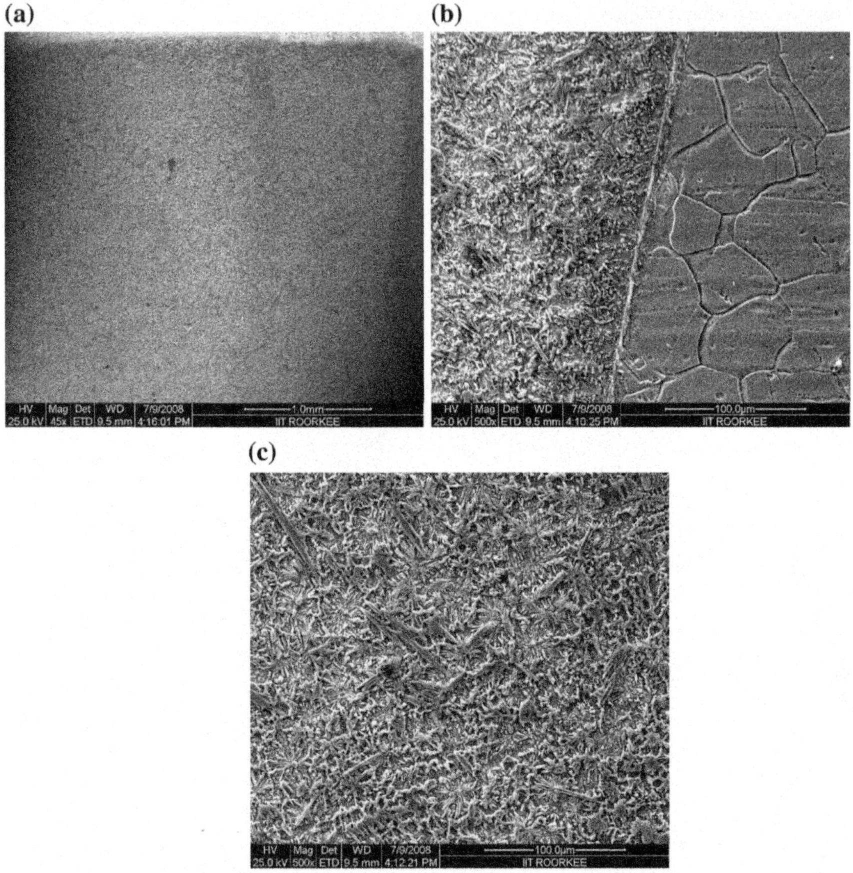

Fig. 6.18 Cladding of Colmonoy 6 by TIG arc **a** SEM macrograph and **b** SEM micrograph of cladding-substrate interface and **c** SEM micrograph of cladding (Kiran et al. 2011; Paul et al. 2014)

Variation in wear rate of TIG claddings of Colmonoy 6 as a function of load and velocity is shown in Fig. 6.19a, b. Wear rate of cladding surfaces increases with increase of load. TIG cladding specimens at 50 N load is subjected to minor seizure-like conditions. Similarly, wear rate increases with increasing velocity. The variation of friction coefficient as a function of load is shown in Fig. 6.20. It is observed that friction coefficient increases with increase in load from 20 to 50 N.

SEM images of worn-out surfaces of Colmonoy 6 of TIG cladding are shown in Fig. 6.21a, b. Oxide wear was not found in TIG cladding specimens. TIG cladding specimens showed wear debris and plastic deformation on the wear surface. Increase in normal load and sliding speed increases damage on the wear surface in the form of microlevel surface layer plastic deformation (Fig. 6.21a, b).

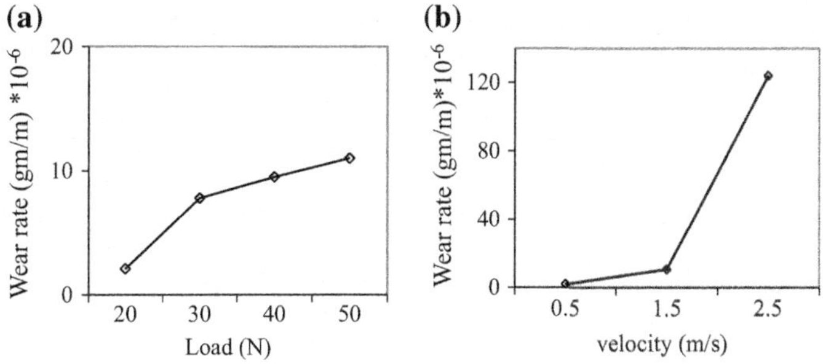

Fig. 6.19 Variation in wear rate as function of **a** load at constant velocity of 2.0 m/s and **b** velocity at constant load of 50 N (Kiran 2009)

Fig. 6.20 Effect of load on friction coefficient Colmonoy 6 cladding (Kiran 2009)

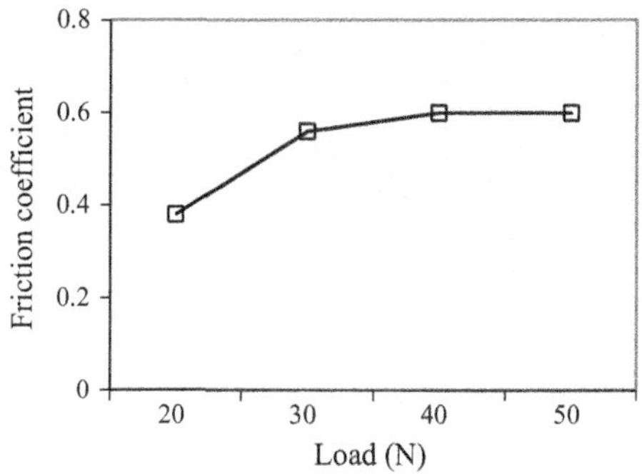

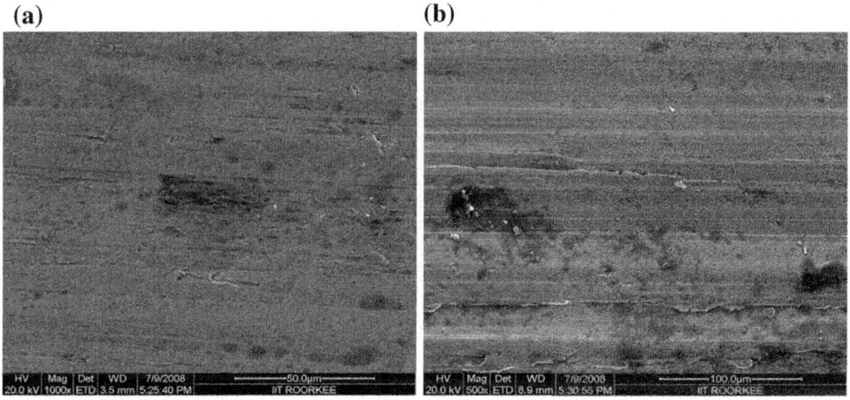

Fig. 6.21 SEM images worn-out surfaces of Colmonoy 6 TIG cladding specimens at **a** 20 N, 2 m/s and **b** 50 N, 2.5 m/s (Kiran 2009)

6.4.7 Plasma Arc Welding

Plasma arc welding (PAW) is similar to the GTAW process except with respect to energy density and penetration capabilities due to higher temperature and more coherent plasma velocity offered by PAW than GTAW. PAW can be considered as an advanced version of GTAW. Like GTAW, PAW also uses the tungsten electrode and inert gases for shielding of the molten weld bead. In PAW, the constriction of arc reduces the cross-sectional area of arc, thereby increasing its energy density (Fig. 6.22). PAW is a high energy density and low heat input welding process; therefore, it poses fewer problems that are associated with weld thermal cycle. The constriction of arc increases the penetration capability and reduces the width of weld bead. The filler for developing weld surfacing is supplied from outside into the plasma zone for melting the electrode and substrate. Plasma velocity, plasma current, and nozzle diameter determine the dilution level and energy density associated with PAW. Therefore, the influence of above parameters of PAW must be considered for their effect on the quality of the welding surfacing. One of the most commonly used variants of plasma arc for weld surfacing is the plasma-transferred arc welding (PTAW) as it allows melting of high-temperature surfacing material, sound metallurgical bonding, and very limited dilution. Substrate becomes a part of electrical circuit in case of PTAW, and arc is established between tungsten electrode and substrate.

Deng et al. (2010) investigated the surface modification of heat resistance steel DIN X45CrSi9-3 with Co-based alloy Stellite 12 using the plasma-transferred arc welding (PTAW). Study revealed three zones namely the fusion zone, the transition zone, and the fine grain zone in the cross section of thick coatings. The microstructure of PTAW Stellite 12 coating can be mainly described as Co-rich α-phase dendritic structure with a network of chromium carbides Cr_7C_3 and $Cr_{23}C_6$ in

Fig. 6.22 Schematic of PAW system showing important components

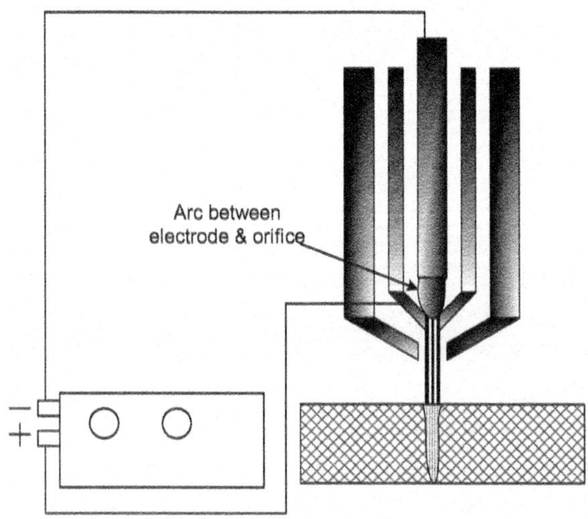

the interdendritic regions. Vickers hardness and bending strength do not degrade much with the loss in coating thickness (from 2.5 to 1 mm) as hardness decreased marginally by about 6%, and the bending strength decreased by approximately 10%. The increase in temperature, lower bending strength by 25%, and the same was attributed to decrease in strength of materials with increasing temperature.

6.5 Laser Cladding

Laser cladding is based on the application of high energy density source of laser beam for melting the material to be used for the development of coating or cladding and substrate. High energy density facilitates fast melting of thin layer of substrate and coating material, which in turn lowers the heat input required for laser cladding (Fig. 6.23). The reduced heat input refines the grain structure (owing to high cooling rate) and significantly lowers the dilution (<1%) and thermal damage of underlying substrate metal. However, the molten pool of weld bead can be protected from atmospheric gases using a shroud of inert shield gas during laser cladding. Coating or cladding materials can either be preplaced on the substrate or fed into the location coaxial, where laser beam is directed for melting the substrate. Extremely high temperature generated by laser and high energy density helps to develop cladding of metals systems having somewhat poor metallurgical compatibility.

Laser cladding may be defined as cladding of a layer of material by melting of preplaced or continuously feed overlay material using laser (as a source of energy)

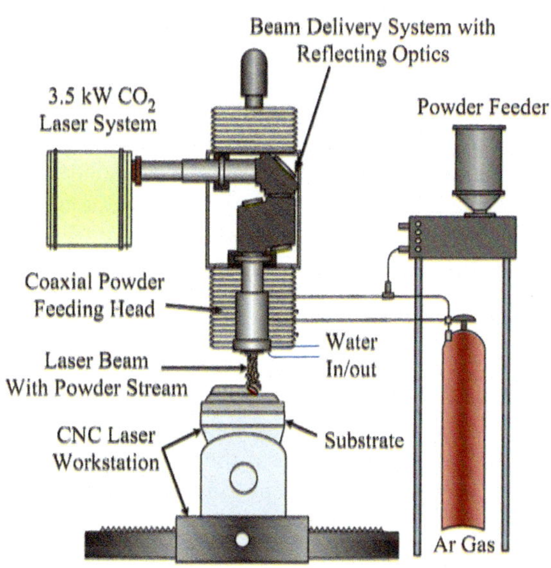

Fig. 6.23 Schematic of laser cladding system showing important components (Kiran 2009)

with limited melting of the underline substrate for producing a metallurgical bond between cladding, and the substrate. Cladding may be used to improve wear, corrosion-resistant, refurbishment of component dimensions and buildup. Laser beam is particularly well suited to accomplish cladding due to the flexibility in directing light energy. Laser cladding offers several advantages over conventional methods:

(a) Dilution is precisely controllable.
(b) Distortion and surface heat affect are minimized.
(c) Fully dense and metallurgical-bonded clad are achieved.
(d) Limited access surface processing is achieved by line of sight.
(e) Selected area is processed which minimizes the wastage of materials.
(f) The process depth is well defined.

Laser cladding process consists of

(i) Metal pool formation and fusion by moving laser beam.
(ii) Supply of cladding material to the substrate.

Cladding deals with conflicting requirement of good metallurgical bonding (by melting of the substrate) and minimum dilution. Laser beam scanning allows well-controlled heat input for the achievement of an acceptable degree of mixing or metallurgical bonding. The methods of supplying cladding material are as follows:

(i) Preplacement of cladding material powder on the substrate; and
(ii) Inert gas propulsion of material powder into a laser generated molten pool.

The relative motion between the laser beam (with powder feeding) and the substrate is used to continuously apply a surface coating. To cover larger surface, the overlapping tracks are made. The powder feed rate must be high enough to build a layer of a coating material on the substrate. The effect of the laser radiation on the material depends on energy density. For avoiding evaporation, laser cladding requires the deposition of a surface layer material between 0.1 and 10.0 mm thick. The choice of laser and relevant parameters affect the microstructure and the properties of the laser-treated material. Rapid solidification starts on moving the laser beam away from the scanned surface. The solidification starts from the bottom of the melt pool substrate interface and progresses toward the surface. The microstructure of the laser clad alloy consists mainly of cast dendrites and HAZ adjacent to the fusion line. The microstructure is affected by the degree of dilution of the substrate.

Study on the effect of laser power and laser scanning speed on geometry, dilution of laser cladding of Stellite 6 and Colmonoy 5 deposited on the austenitic stainless steel (AISI 316L) showed that increase of laser power (a) increases the clad width and dilution level and (b) decreases the clad hight irrespective of laser cladding speed and cladding material (Fig. 6.24a, b). On contrary, increase of laser scanning speed (a) decreases the clad width and clad height and (b) decreases dilution of cladding irrespective of laser power and cladding material.

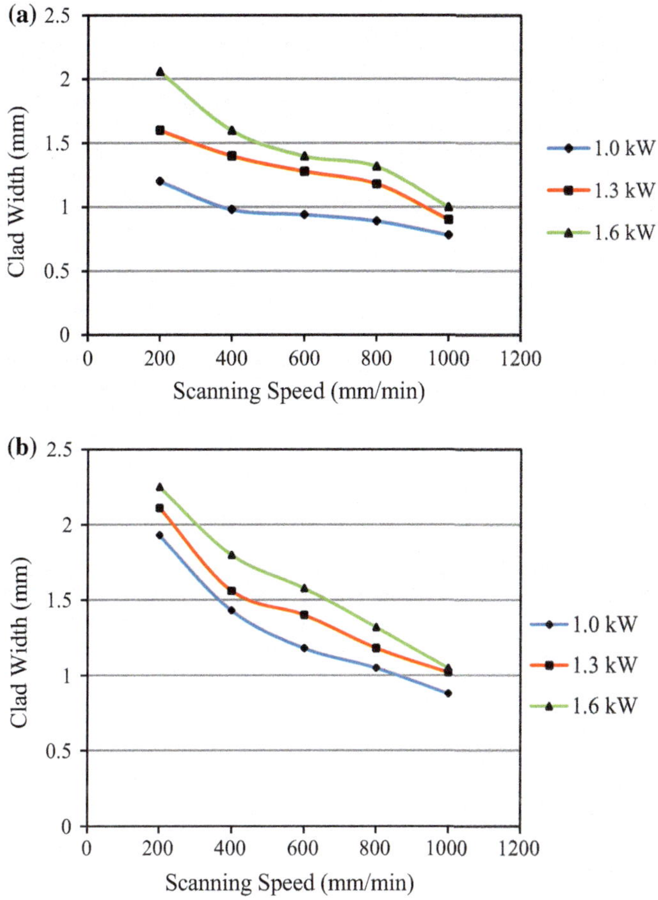

Fig. 6.24 Influence of scanning speed and laser power on the clad width of Stellite 6 and Colmonoy 5 cladding (Kiran 2009)

Laser clad surface should be free from defects like cracks, porosity near interface, and good bonding between the substrate material (Fig. 6.25a). Typical microstructure of cladding of Colmonoy 5 shows a columnar dendritic growth near substrate–cladding interface. The microstructure exhibits primarily dendrite of Ni solid solution and eutectic mixture in the interdendrite region (Fig. 6.25b). Uniform dendrite growth is evident in the direction of buildup surface.

Microstructural studies of Stellite cladding by laser showed good bonding of the clad layer to the substrate (Fig. 6.26). Stellite cladding on AISI 304L substrate shows dendrite and inter-dendritic eutectic. The basic microstructure of Stellite laser clad reveals hypoeutectic structure, containing fine and homogeneous dendrites and the interdendritic eutectics. According to the Co–Cr–W phase diagram, a Co-rich

(a) (b)

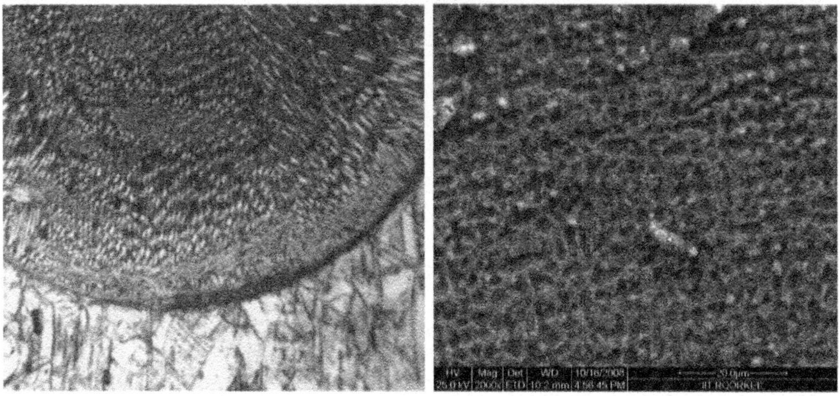

Fig. 6.25 Micrograph of laser cladding **a** optical micrograph and **b** SEM micrograph (Kiran 2009; Kiran et al. 2011)

phase is first formed dendritically from the liquid state. Cr and C become enriched in the remaining liquid in the interdendrite regions and eutectic carbide structure forms. Typically, the dendrites consist of the cobalt-based solid solution and are surrounded by eutectics composed of M_7C_3-type carbide (M = Cr, Co, W) and the cobalt-based solid solution.

Microhardness variation of laser cladding of Colmonoy 5 and Stellite 6 shows significant increase in hardness on approaching from stainless steel substrate to laser cladding (Fig. 6.27a, b). Transition in hardness from substrate to cladding is gradual and can be attributed to dilution of cladding material on mixing with substrate. Moreover, the extent of the dilution is limited to about 60–70 μm (Fig. 6.28).

The erosion behavior Colmonoy 5, Stellite 6 cladding, and AISI 304L steel substrate as function of orientation angle during slurry erosion test is presented in Fig. 6.29. Colmonoy 5 cladding enhances erosion resistance in the range of 41–62%. The Stellite 6 clad layer also shows similar trend of the erosion, with maximum wear at 30° orientation angle. Stellite 6 clad shows better erosion resistance in the range of 49–75% compared to the substrate material, i.e., AISI 304L steel over the range of orientation angles. A possible reason for better erosion resistance of the Stellite 6 clad layer may be the lower degree of dilution and higher in microhardness of the clad layer than substrate and Colmonoy 5.

SEM study of the Colmonoy 6 and Stellite 6 clad surfaces (Figs. 6.29 and 6.30) shows few small lipless craters after erosion at 45° and 60° orientation angles. This can be attributed to the difference in the microhardness value of the clad surfaces than the substrate. Further, increase in the orientation angle increases erosion rate. The clad surface oriented at 15° angle depicts deeper and larger impact craters compared to those on the clad surface at the orientation angles of 30° and 45°

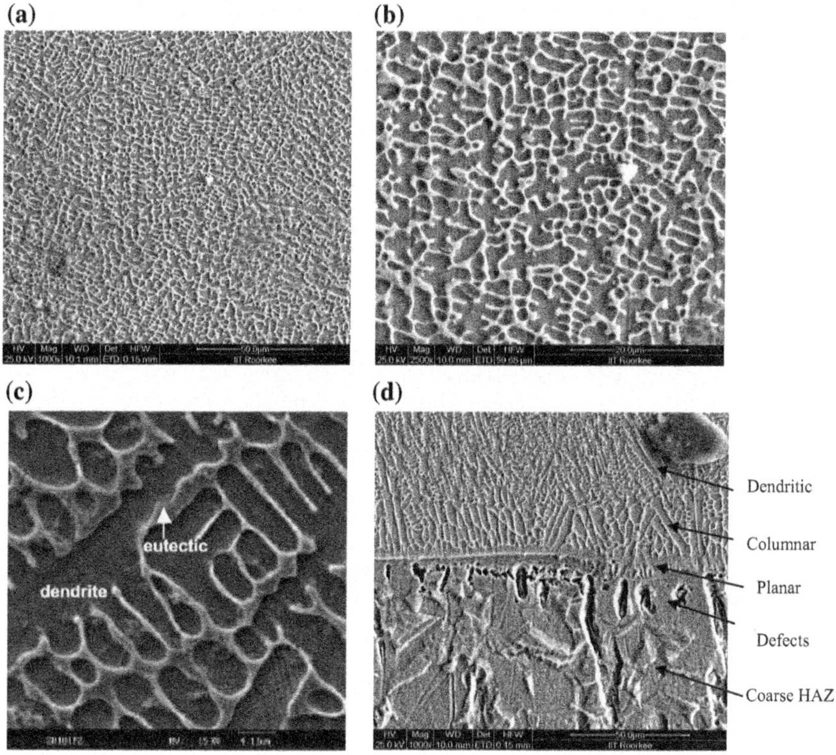

Fig. 6.26 SEM images of Stellite 6 claddings on AISI 304 L substrate material **a–c** micrograph of cladding at increasing magnification and **d** cladding-substrate interface (Kiran 2009)

angles. These features in turn result in higher material removal from the clad surface. At acute angles, the normal component of particle velocity is very small which may not be sufficient to penetrate at the clad surface or to deform it plastically.

6.6 Thermal Spraying

All thermal spray processes work on the common principle associated with the developing of thin films and coatings. The feedstock, that is, the material to be coated (in form of wire, powder) is passed through suitable heat source such as flame, arc, plasma. and brought to plastic or complete and/or partial molten state. Depending upon the heat source, the temperature can vary significantly. For example, in the oxy-fuel combustion flame (flame spraying, HVOF, cold spray,

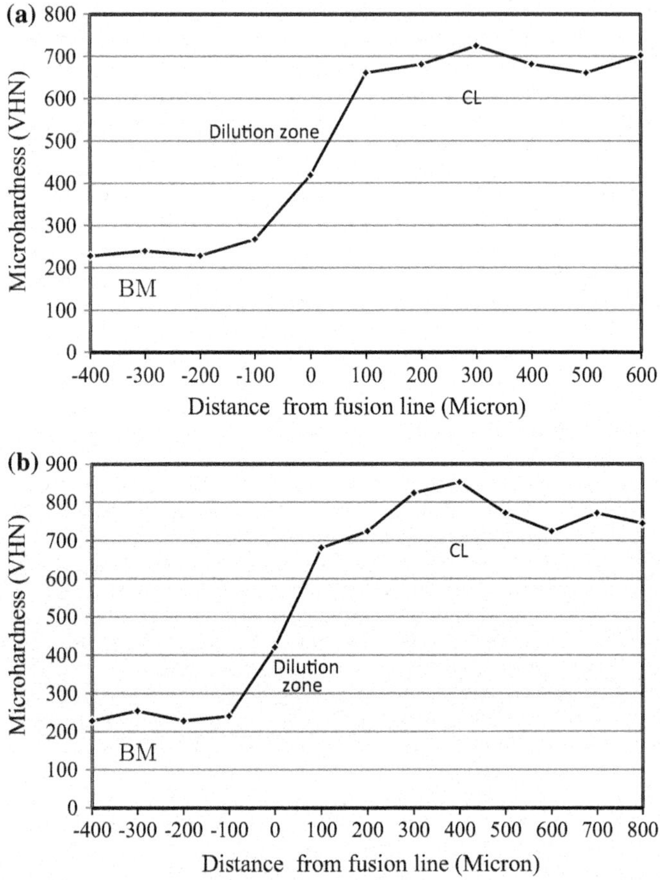

Fig. 6.27 Microhardness profile of laser claddings of on stainless steel substrate of **a** Stellite 6 and **b** Colmonoy 5 deposited using laser power of 1.6 kW at scanning speed of 800 mm/min (Kiran 2009)

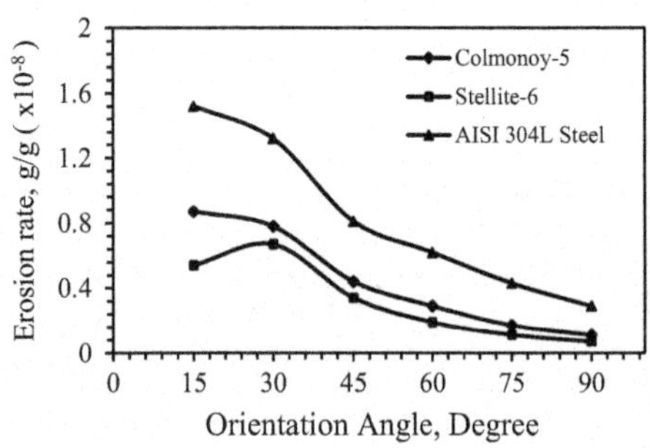

Fig. 6.28 Variation of erosion rate of Colmonoy 5, Stellite 6 cladding surfaces and AISI 304 steel substrate with orientation angle (Kiran 2009)

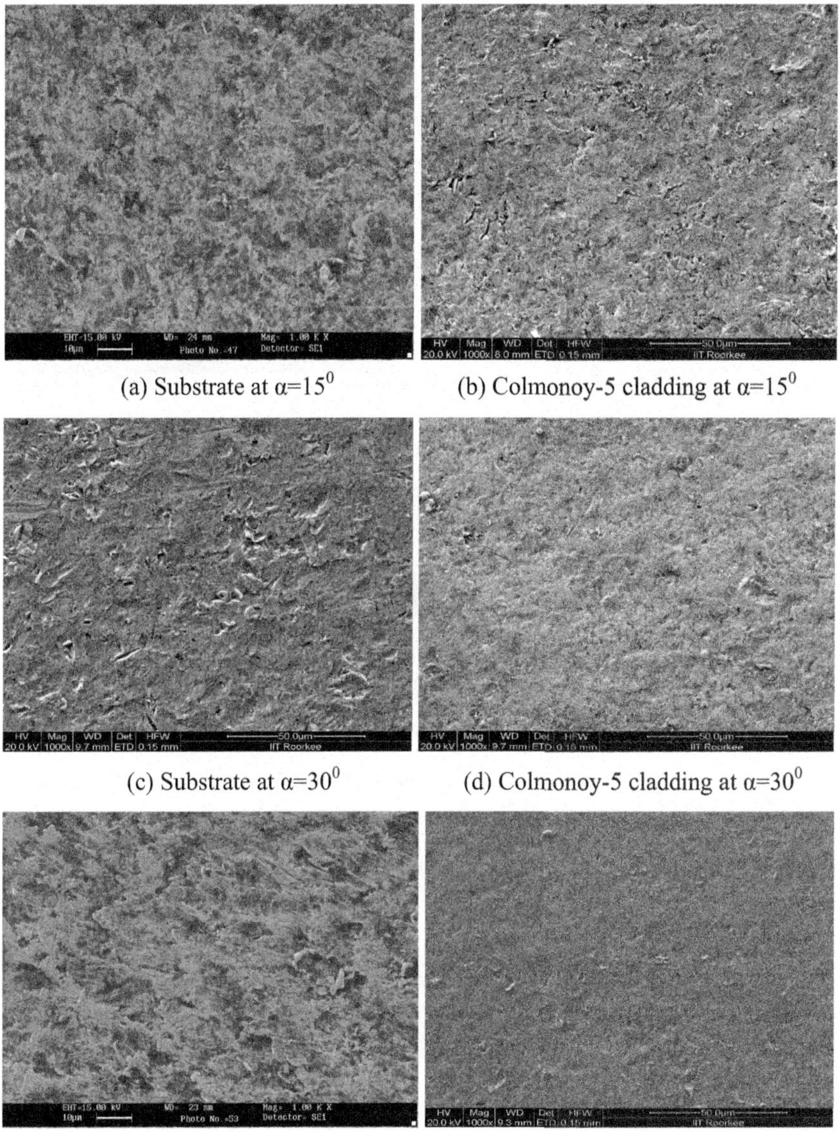

(a) Substrate at α=15⁰ (b) Colmonoy-5 cladding at α=15⁰

(c) Substrate at α=30⁰ (d) Colmonoy-5 cladding at α=30⁰

(e) Substrate at α=45⁰ (f) Colmonoy-5 cladding at α=45⁰

Fig. 6.29 Mechanism of material removal for AISI 304L steel and Colmonoy 5 cladded surfaces at oblique impact angle (d = 550 μm, C_w = 20% and V = 3 m/s)

detonation spray) process, the maximum temperature is around 3000 °C and arc temperature (in case of arc spray) is about 6000 °C, and plasma gas (in plasma spray process) offers further higher temperature, which is approximately 10,000–15,000 °C.

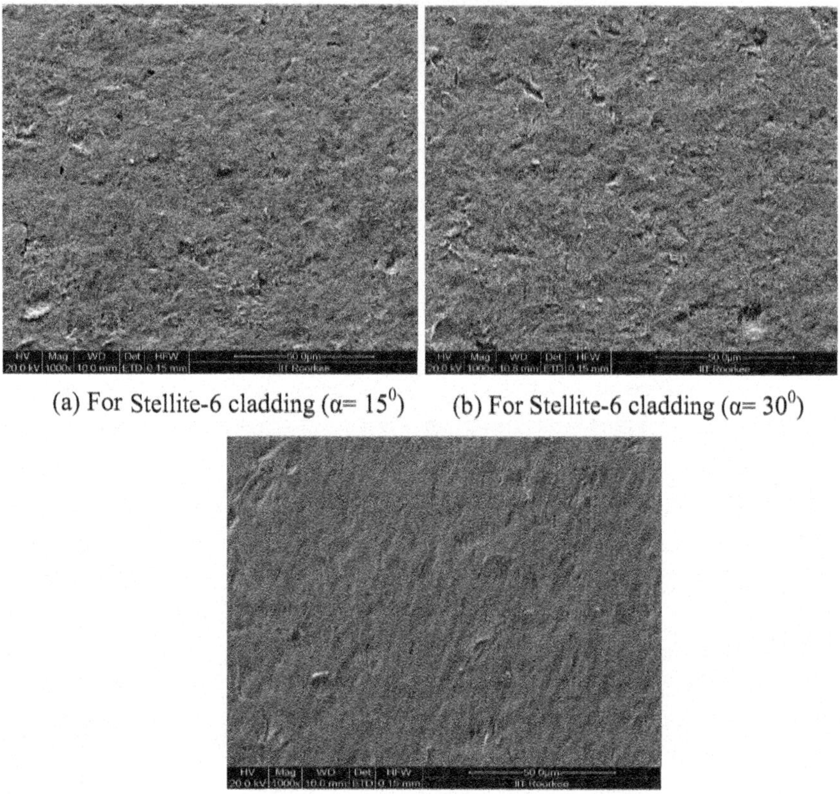

(a) For Stellite-6 cladding ($\alpha = 15^0$) (b) For Stellite-6 cladding ($\alpha = 30^0$)

(c) For Stellite-6 cladding ($\alpha = 60^0$)

Fig. 6.30 Mechanism of material removal for Stellite 6 cladded surface at oblique impact angle (d = 550 μm, C_w = 20% and V = 3 m/s) (Kiran 2009)

During their flight in heat source, the solid, molten or semi-molten particle is accelerated and allowed to impinge with substrate. The period of stay of the feedstock in the heat source is determined by their velocity. The velocity can vary over a wide range from 50 to 3000 m/s. Accordingly, thermal damage of the hard and stable particles (like tungsten carbide, chromium carbide that are not expected to overheated or melting) will be occurring. Higher the velocity, lower will be the thermal damage and interactions between the gases (around the particle) and molten or semi-molten feedstock, which can lead to the formation of inclusions. The velocity and mass of accelerating particles affect gas and feedstock interactions, stay of feedstock at high temperature in the heat source region, the kinetic energy of (solid, molten, or semi-molten) particles impinging with the substrate surface. In general, an increase in the kinetic energy of impinging particles increases the bond strength of substrate and lowers the porosity due to better metallic intimacy between the molten particle forming splats after impact or already deposited splats.

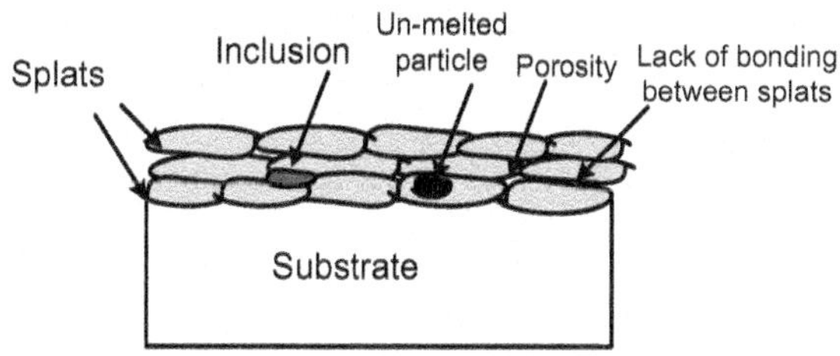

Fig. 6.31 Schematic of thermal spray coating structure

Low velocity or kinetic energy can result in pores at the coating–substrate interface and between splats being deposited. Based on the heat source, thermal spray process has been given different names such as spray and fuse, flame spray, high-velocity oxy-fuel spray, detonation spray, plasma spray, electric arc spray.

The basic component in the formation of a coating is the "splat". "Splat" is the term given to a single impacted droplet or particle. Many overlapping splats solidify and adhere to form a continuous layer. Splats are created when the accelerated and molten particles impact on a prepared surface. The arriving molten droplets are generally spherical and based on the impact of the substrate surface they spread over and fill the underlying interstices (spaces). The droplet becomes flattened, and disk-like structure, which when polished in the cross section, looks like the coating microstructures as shown in Fig. 6.31.

Thermal spray processes are also characterized by rapid solidification. As the relatively small individual particles impact the more massive substrate, their heat is liberated quickly. Solidification rates for metals are in the range of 10^5–10^8 °C/s and such rapid cooling rates produce a wide range of material states from amorphous to metastable. Two structures are generally present within the coating of splat structure and intersplat structures. Within the splat, the rapid solidification of thermal spray creates grain size less than 5 μm and prevents many materials from achieving their equilibrium phases, resulting in anisotropic coating properties. In most of the thermal spray coatings, the properties in the *x*- and *y*-planes, that is, the plane parallel to the substrate surface are same. However, in the *z*-plane, which is a perpendicular to the substrate surface, the properties vary significantly from those in the *x*- and *y*-planes. Despite the possibilities of oxide inclusions in the coating, the rapid thermal processing and solidification in the thermal spray coating lead to compositional homogeneity, which is a representative of the starting feedstock compositions.

6.6.1 Flame Spraying Process

The heat required for heating the feedstock in the form of mostly powder or wire is obtained from the combustion of oxygen-fuel gas (acetylene). The maximum temperature generated by combustion of oxy-acetylene flame is about 3000 °C, and the velocity is about 50 m/s. The fed stock powders or wires are introduced through

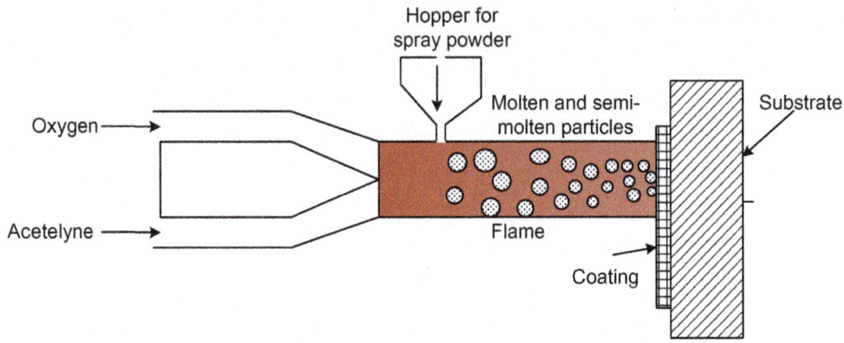

Fig. 6.32 Schemtic of oxy-acetelune flame spray system

the nozzle into the flame. The feedstock mainly in powder form is melted, and the molten or semi-molten droplets are accelerated toward the substrate surface (Fig. 6.32). The flame spray process offers many advantages such as low capital cost, high deposition rate and low wastage of feedstock, and relative ease of operation. The flame spray process develops coating with 5–10% porosity, and in accordance to the porosity, the density varies from 85–98% depending upon the hardness of coating material. However, the flame spray process owing to low velocity and high heat input suffers from coarse grain structure of coating with comparatively high percentage of porosity and inclusion.

Rambabu and Dwivedi (2006) studied slurry erosion behavior of flame spray coating of Ni–WC deposited on the mild steel substrate. The coating was reported to reveal mainly polyhedral shape primary tungsten carbides particles in the matrix eutectic–carbides. Erosion of the coating increased with increase of the rotational speed irrespective of silica sand slurry concentration (Fig. 6.33). Coating however exhibited more erosion with increase of slurry concentration. Development of nickel–tungsten coating improved the erosion resistance of mild steel substrate by

Fig. 6.33 Effect of rotational speed on erosive wear of substrate and coating under different sand slurry concentration (Arji 2006)

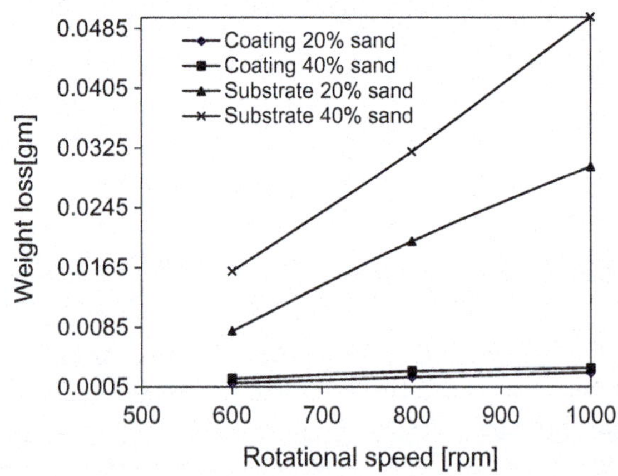

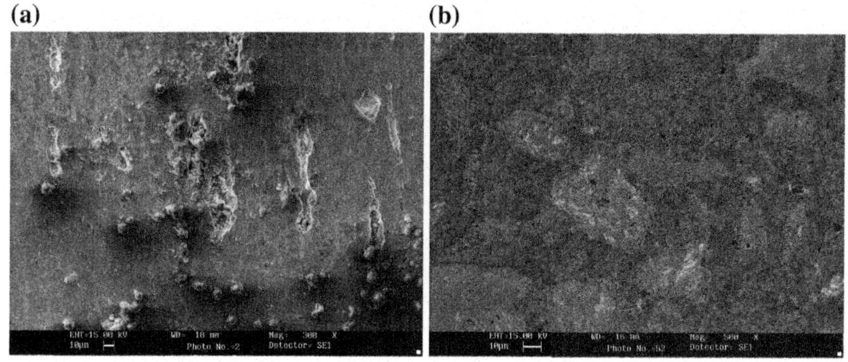

Fig. 6.34 SEM images of erosion wear surface a mild steel substrate and b nickel–tungsten carbide coating subjected erosion in 20% sand slurry at 600 rpm (Arji 2006)

8–16 folds as per erosion wear conditions. The operational erosive wear mechanisms were pitting, crater formation, and scratching. Pitting was observed on nickel-rich eutectic matrix, while scratches and indents were found on the surface of tungsten carbide particles (Fig. 6.34).

Harsha et al. (2006) investigated the abrasive wear behavior of Ni–WC, Ni–Cr–B–Si–C, and Stellite coatings deposited by flame spraying on mild steel substrate. Substrate mainly exhibited ferrite and pearlite in microstructure (Fig. 6.35a) Microstructure showed different microconstituents in each of above coating (Fig. 6.35b–d). Ni–WC coating revealed the WC particles in the eutectic matrix. Ni–Cr–Si–B coating showed the cellular structure of nickel solid solution and low melting point phase segregated along the grain boundaries. Co–Cr–W–Ni–C coating showed primary cuboid-shaped Cr-based carbide particle in the matrix of eutectic. The hardness of Ni–WC coating was higher than Co–Cr–W–Ni–C and Ni–Cr–Si–B coatings. Lowest hardness was found in the case of Ni–Cr–Si–B coating. Wear rate of all coatings investigated has been found to increase with increase in normal load. Wear rate of Ni–WC coating was lowest and that for Ni–Cr coating was found maximum (Fig. 6.36). SEM study showed that wear of Ni–Cr–Si–B coating takes place by ploughing and microgroove formation, while in case of Ni–WC coating active wear mechanism is fragmentation of WC particles and scratching of eutectic matrix (Fig. 6.37).

Microhardness of primary tungsten carbide particle (3520–6150 HV) was found higher than eutectic matrix (910–1460 HV) in nickel–tungsten carbide coating. Further, average hardness of primary tungsten carbide and eutectic was 5030 and 1220 HV, respectively. Average hardness of nickel solid solution in Ni–Cr–Si–B coating was 280 VHN, while that of eutectic was found in range of 380–430 HV. Average microhardness of Co–Cr–W–Ni–C coating was 480–640 VHN.

Wear of Ni–WC coating took place by scoring of eutectic matrix and fragmentation of carbide particles (Fig. 6.37a). Wear of Ni–Cr coating occurred largely by groove formation, microcutting, plowing and scoring of matrix (Fig. 6.37b).

Fig. 6.35 Microstructure of **a** steel substrate, **b** nickel–tungsten carbide, **c** Ni–Cr–Si–B, and **d** Co–Cr–W–Ni–C coating (Harsha et al. 2006)

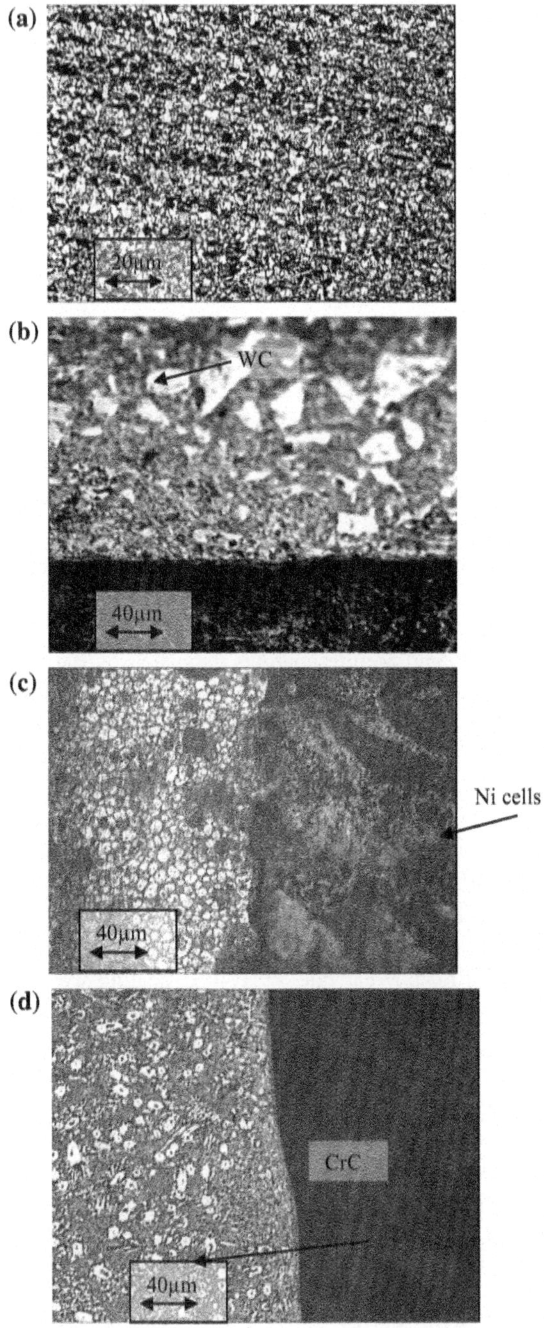

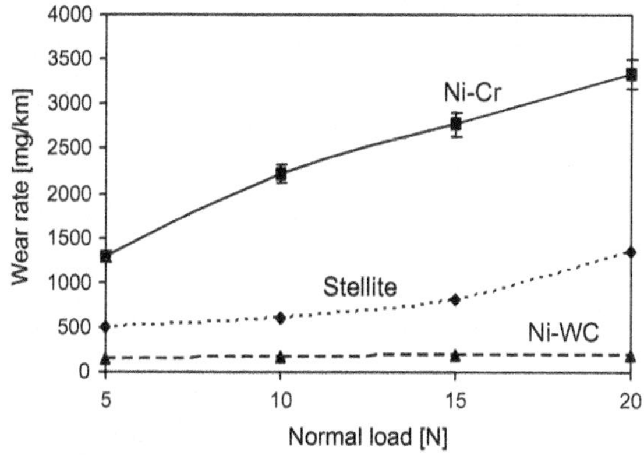

Fig. 6.36 Wear rate versus normal load relationship for different coatings against 120 grade abrasive medium (Harsha et al. 2006)

Damage on the wear surface of Co-based alloy coating showed microgrooves and ploughing (Fig. 6.37c). Plastic deformation can be seen as extruded fins at the edge of grooves produced by rubbing. These fins can get detached to form secondary chips when coalescence of cracks takes place. It is generally believed that when abrasives come in contact of the wear surface, two processes take place (1) formation of plastically impressed groves which do not involve metal removal (2) the separation of metal particles in form of chips. Abrasives will be cutting/extruding the fins (deforming plastically) which depends upon the shape of indenting abrasive particles.

Prerequisite for abrasive wear is penetration by hardness abrasive particle because without penetration there won't be any loss of material under such conditions. Higher the hardness greater will be the abrasive wear resistance. Experimental wear results are in line of this fact. Maximum hardness was found in case of Ni–WC coating, and it showed the lowest wear rate (best wear resistance). Lowest hardness was shown by Ni–Cr–Si–B coating, and this coating was subjected to maximum wear rate (Fig. 6.36). SEM study also supported the wear results. Ni–Cr–Si–B coating was subjected to plowing (owing to low hardness), while Ni–WC coating showed microcracking and fragmentation of hard coating. Further, owing to relative motion between the abrasive particle and wear surface, scratches are formed. Cross section of these scratches is governed by various material-related characteristics and test conditions. Material parameters such as hardness, ductility, flow stress, brittleness, work hardening characteristics, and microstructural aspects (size, shape, distribution of various phases, and their relative amounts) and test conditions such as size and shape of abrasive, hardness of abrasive particles, relative speed between abrasive and wear surface, load being applied on the wear surface through abrasives, etc., should be considered for analyzing the wear behavior.

Scratch formed due to abrasive action may or may not remove material. Even deep penetration by abrasives in soft and ductile material may not remove the material equal to the volume of scratches that are formed on the wear surface due to

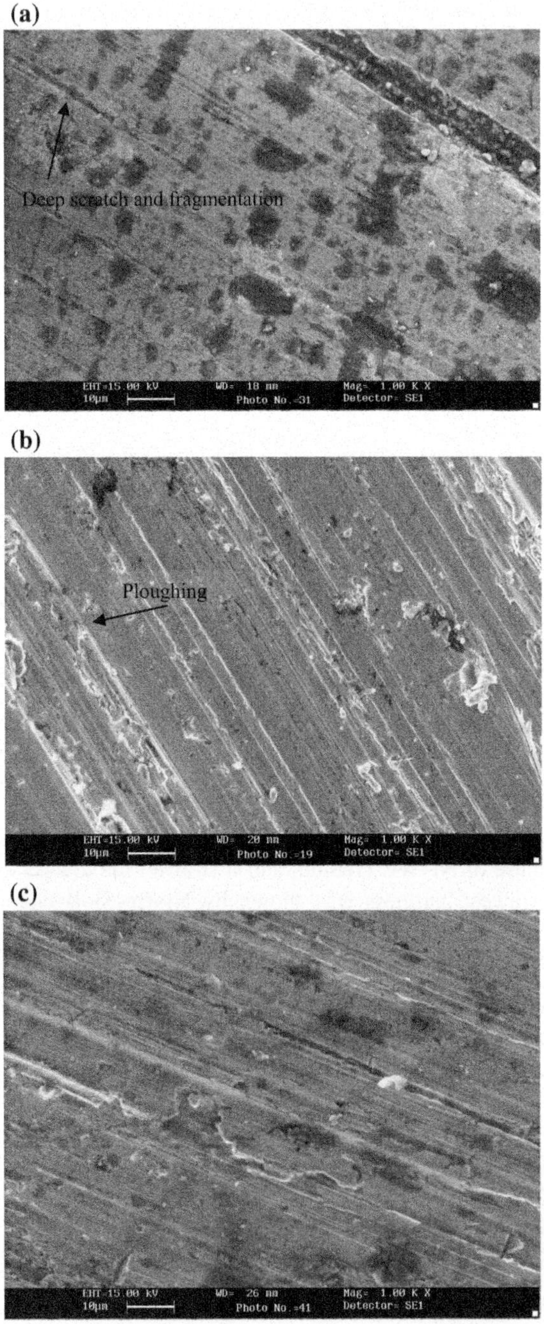

Fig. 6.37 SEM images of wear surface of **a** Ni–WC, **b** Ni–Cr–Si–B, and **c** Co–Cr–Ni–W–C coating after sliding at 20 N load against 120 grade abrasive medium (Harsha et al. 2006)

easy plastic flow of metal in direction perpendicular to that of sliding. This sidewise flow leads to formation of ridges on both sides of groove. Repeated deformation of material of these ridges work hardens them. Work-hardened materials under stress system developed during the sliding result in cracks, and eventually coalescence of these cracks results in detachment of materials from the wear surface as wear debris. Hard and brittle coating materials do not get deformed to a large extent to form these ridges. Further, these materials develop cracks easily under the influence of stress developed during sliding due to lmited ability to get deform. Coalescence of these cracks produces small metallic pieces as debris.

Improvement in performance of flame spray coating

The flame spraying is considered as a (a) low particle velocity, (b) high-temperature exposure, and (c) prolonged inflight exposure of coating material during spraying. These features in turn lead to (a) low coating–substrate bond strength, (b) high porosity, (c) increased thermal degradation of coating material, and (d) coarse grain structure. Moreover, flame spraying is still popular in industries owing to ease of handling, ease of implementation in manufacturing and economical. Therefore, attempts have been made to overcome some of the issues related to flame spray using (a) densification-based methods such as hot isostatic pressing, rolling, compaction, (b) re-melting using higher energy density heat source like GTAW and laser, (c) grain refinement using controlled alloying of rare earths such as ceria and lanthanum oxides. Application of compressive force at high temperature for densification increases (a) density, and (b) bonding coating–substrate interface.

Compaction of flame spray coating using roller

The compaction of the flame sprayed coatings helps to increase the soundness and density of the coatings. The compact using 200 N load applied through a roller made of steel (25 mm diameter and 30 mm length) using a system is schematically shown in Fig. 6.38. The movement of roller is synchronized with flame spray torch. Sticking tendency of roller with coating is reduced by pouring silicone lubricating oil. The compaction reduces the average grain size of nickel grains from 28 to 18 μm (Fig. 6.39), and the same is attributed to dynamic recrystallization due to compaction at elevated temperature. The grain structure of compacted coating at middle was finer than at top and near interface. Apart from grain refinement, the porosity of the coating after the compaction reduced by 32% and it was attributed to collapse of gaseous and shrinkage pores under the influences of pressure at elevated temperature during compaction. The reduction in porosity and refinement of grain structure compacted coating in turn increased the hardness 194–233 HV_5.

Re-melting of flame sprayed coating

Attempts have been made to improve the structure, hardness and wear resistance of flame sprayed coating (Dhakar and Dwivedi 2010). Authors reported the flame spraying of Ni–Cr–B–Si coating using neutral flame of oxy-acetylene gas at 0.3 and 0.12 MPa pressures of oxygen and acetylene, respectively on the mild steel substrate preheated to 200 ± 10 °C. The flame sprayed coating in as-sprayed condition showed coarser structure with little pores. Re-melting of coating using a TIG welding

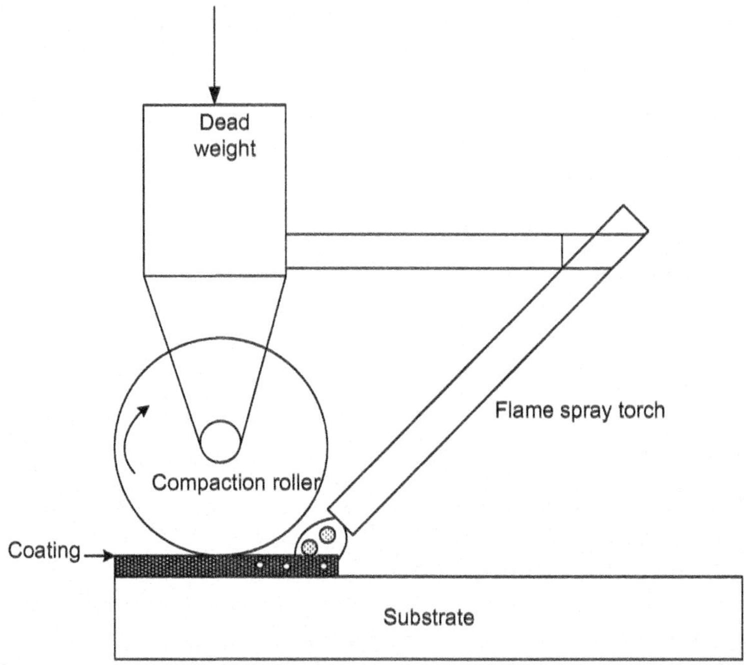

Fig. 6.38 Scheme of continuous compaction during flame spraying

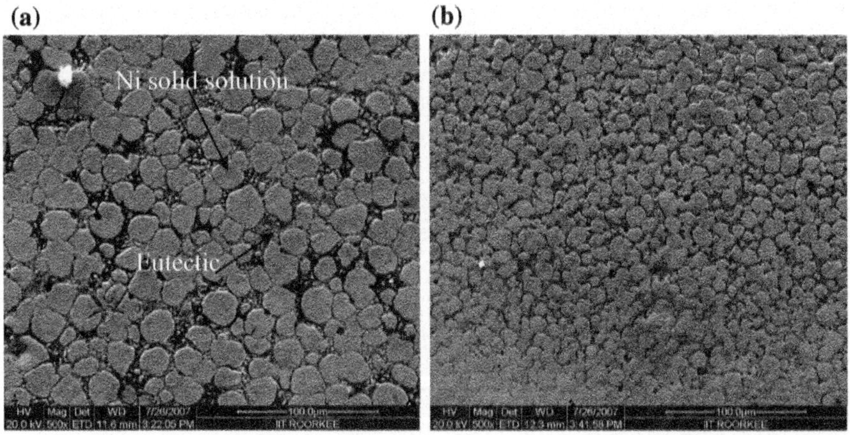

Fig. 6.39 Micrographs of flame sprayed Ni–Cr–B–Si coating in **a** as-sprayed and **b** compacted condition (Sharma et al. 2008, 2010)

arc heat (torch at welding current of 140–180 A and arc voltage 16–20 V) refines the grain structure. The microstructures of the coatings in the as-sprayed and re-melted conditions are shown in Fig. 6.40a, b respectively. The coating mainly contains cellular grains of Ni solid solution in eutectic matrix. The optical microscope image

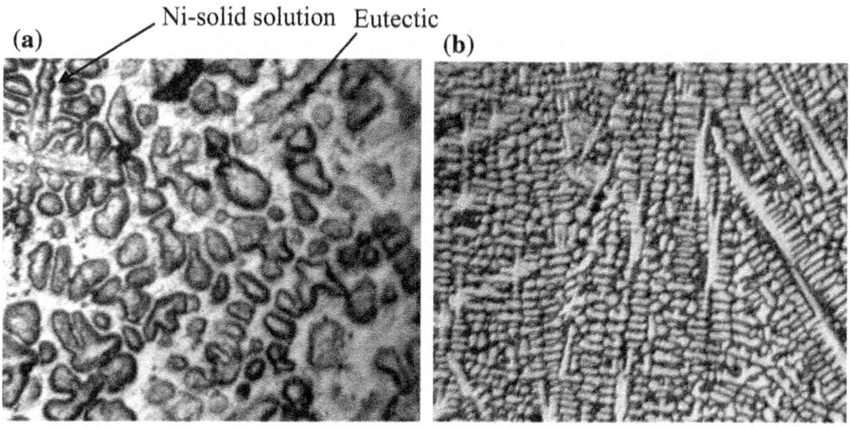

Fig. 6.40 Microstructures of the flame sprayed Ni–Cr–B–Si coating in **a** as-sprayed and **b** re-melted condition (Dhakar et al. 2012)

analysis of the coatings exhibits mainly two types of microconstituents, namely Ni solid solution (dark) and fine eutectic (bright). Image analysis of these coatings revealed that fraction of Ni solid solution is 52.7% and eutectic is 47.3% in the as-sprayed conditions, while in the re-melted conditions. these are 54.7 and 45.3% respectively. Re-melting of coating reduces average grain size of Ni solid solution from 39 to 32 μm. Refinement of grain structure is attributed to higher cooling rate experienced by coating during solidification than cooling rate experienced during flame spraying. Re-melting also increases the hardness from 318 to 352 HV and abrasive wear resistance of flame sprayed coating irrespective of load and abrasive grit size (Fig. 6.41).

Fig. 6.41 Abrasive wear behavior of Ni–Cr–B–Si coating in as-sprayed and re-melted condition against different loads and abrasive grit sizes (Dhakar et al. 2012)

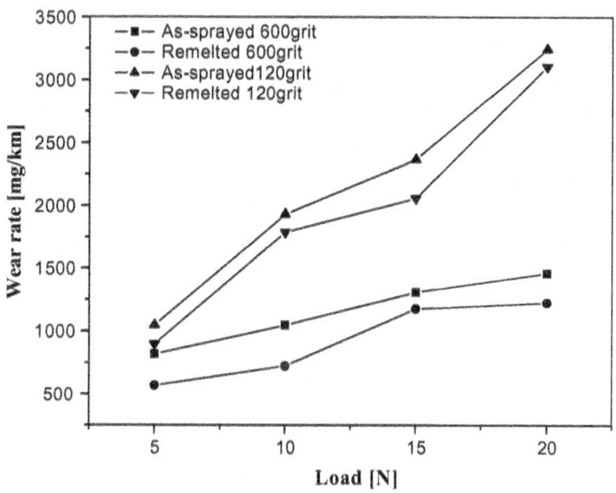

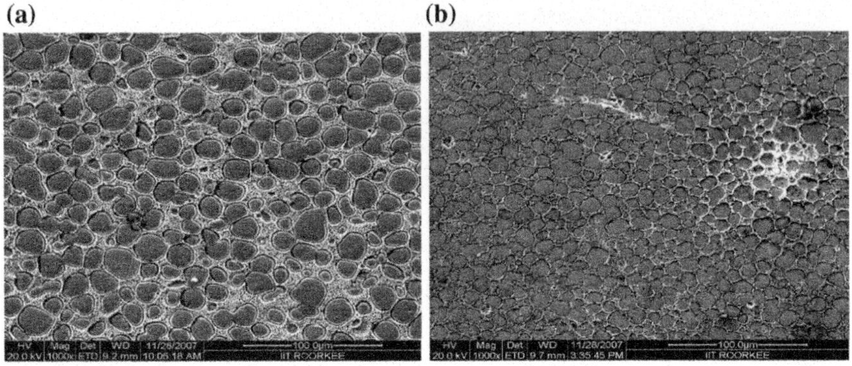

Fig. 6.42 Microstructure of flame sprayed condition in **a** as-sprayed conditions and **b** 1.2% La-modified coatings (Sharma et al. 2010)

Controlled alloying using rare earth

Some of rare earth elements act as heterogeneous nucleants. Addition of these nucleants with coating material in the powder form helps to refine grain structure of the flame sprayed coating. Effect of La_2O_3 addition on microstructure, hardness and abrasive wear resistance of flame sprayed Ni–Cr–B–Si coatings revealed that the addition of La_2O_3 in the NiCrBSi coatings refines the microstructure (Sharma and Dwivedi 2009). The optimum addition of (1.2%) La_2O_3 reduces the grain size of Ni solid solution cells from 25 to 15 μm (Fig. 6.42a, b). The optimal addition of La_2O_3 distributes the various elements more uniformly as compared to as-sprayed coating. The optimal addition of La_2O_3 in the NiCrBSi flame sprayed coatings increases the hardness of the coating by 1.2 fold. Abrasive wear rate of the coatings with optimum La_2O_3 (1.2 wt%. La_2O_3) is 25% lower than the as-sprayed coatings. The increase in abrasive wear resistance of the optimum La-modified coating was attributed to grain refinement and increase in hardness.

Purkayastha and Dwivedi (2014) reported that the addition of (0.8 wt%) CeO_2 reduces the grain size of Ni solid solution cells from 36 to 23 μm. The addition of CeO_2 in flame-sprayed modified NiCrBSi coatings improves the microstructure in terms of grain refinement. The addition of CeO_2 improves the average microhardness of the coating by 32%, and the same is attributed to grain refinement. The abrasive wear rate of the coatings containing 0.8% CeO_2 is 37% lower than those without CeO_2 (Fig. 6.43a–d). The increase in the abrasive wear resistance is due to grain refinement and microhardness of the coatings.

Ceria addition (0.9 wt%) in flame sprayed Ni–WC coating increases hardness and abrasive wear resistance and reduces porosity. The improvement in hardness is attributed to the lower porosity levels in modified coatings and the enrichment of the Ni alloy matrix with carbon and tungsten due to accelerated dissolution of WC by addition of CeO_2. The improvement in tribological property of CeO_2-modified coating is more at high load than low load. The CeO_2-modified coatings, especially 0.6 and 0.9 wt% CeO_2 exhibit better anti-wear properties primarily due to excellent

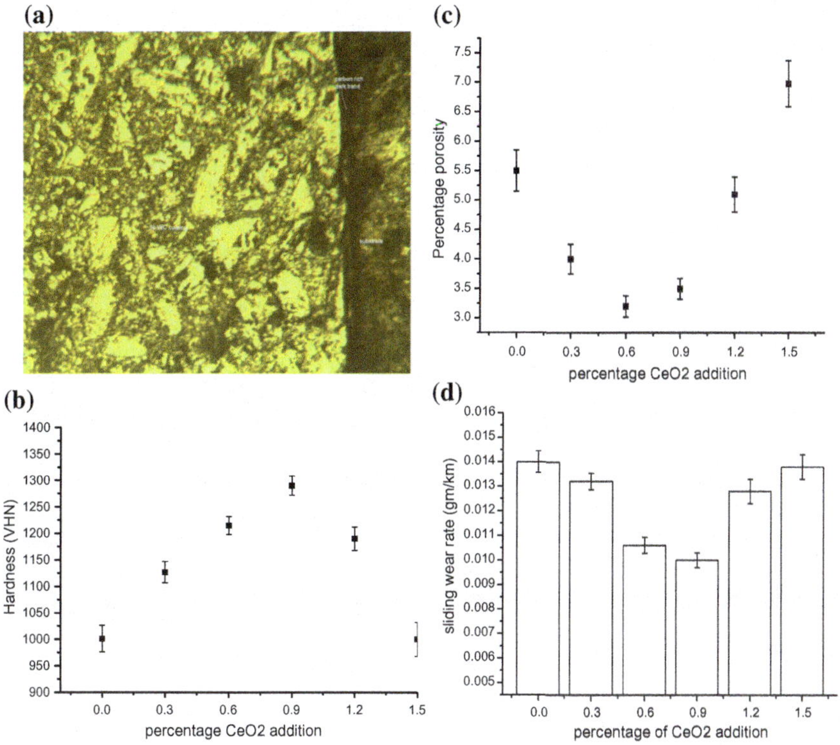

Fig. 6.43 Effect of ceria addition in Ni–WC flame sprayed coating on **a** microstructure, **b** hardness, **c** porosity, and **d** sliding wear at 1.5 m/s under 50 N load (Purkayastha 2014)

lubrication function of CeO_2 at high temperatures because of its hexagonal layer structure and increased hardness of the binder, lower porosity levels, and improved hardness.

Reinforcement of hard tungsten and chromium carbide particle

Incorporation of hard carbides (chromium carbide, tungsten carbide), borides in matrix of comparatively soft flame sprayed coating materials increases the hardness and wear resistance under adhesive and abrasive wear conditions. The hard particles make the coating strong, coherent, and stable against external load by providing support to matrix materials. Additionally, such particles break the continuity of abrasive grove and even fracture of abrasive particle causing wear. Harsha and Dwivedi (2008) investigated the effect of addition of CrC particles in Ni–Cr–B–Si flame sprayed coating on the structure and properties (Fig. 6.44a–d). Microstructure of the as-sprayed coating exhibited honeycomb-type cellular structure consisting nickel solid solution surrounded by eutectic network. CrC-modified coating in as-sprayed condition showed CrC particles in the matrix while after heat treatment these particles get disappeared (Fig. 6.44d). This Ni–Cr–B–Si is designated as 1004

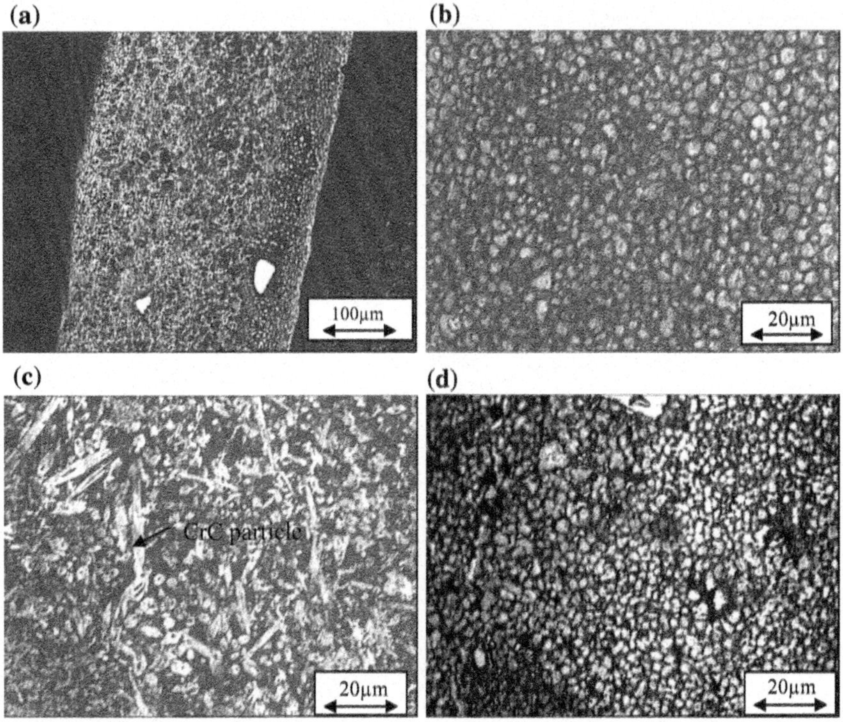

Fig. 6.44 Optical microphotograph of Ni–Cr–B–Si coating **a** overview of coating, **b** microstructure of unmodified as-sprayed coating, **c** coating with 20%CrC addition in as-sprayed condition, and **d** coating with 20%CrC addition in heat-treated condition (Harsha et al. 2008)

in legends of Fig. 6.45. Increase in microhardness of the coating was found with addition of CrC particles (Fig. 6.45a). Probably post-spray heat treatment dissolves CrC particles in the matrix. In general, increase of normal load increased the abrasive wear rate of both as-sprayed and CrC-modified coating. Addition of CrC particles lowers the wear rate of modified powder coating 3–8 fold compared to that of unmodified coating. Addition of carbide particles in unmodified powder coating is more beneficial against coarse abrasive than fine abrasive (Fig. 6.45b). Loss of material from the wear surface of the coating takes place primarily by crater formation and ploughing mechanisms.

6.6.2 High-Velocity Oxy-Fuel Thermal Spraying

High-velocity oxy-fuel (HVOF) thermal spraying is an advanced version of flame spray process, where the combustion of oxygen and fuel gas mixture generates the heat and leads to temperature about 3000 °C. The feedstock powder or wire

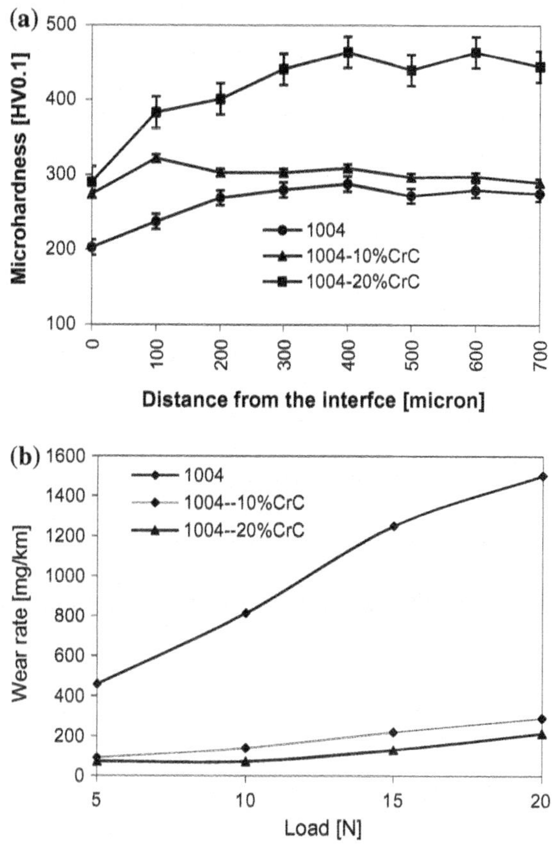

Fig. 6.45 Effect of CrC addition on **a** microhardness and **b** abrasive wear behavior of coatings against 600 grade abrasive medium (Harsha et al. 2008)

material is melted in the combustion zone and then propelled at a high velocity (up to 1000 m/s) toward the surface. In the combustion zone, feedstock is brought to molten or semi-molten state depending on the melting temperature and the feed rate of the material. The quality of the coatings is predominantly determined by velocity and temperature of the molten, semi-molten, softened particles impacting on to the substrate. The common fuel used in HVOF process includes C_3H_6, C_3H_8, H_2, C_2H_2, CH_4, LPG, C_2H_4, kerosene, liquefied petroleum gas (LPG), and methylacetylene-propadienestabilized gas. Components of HVOF system are schematically shown in Fig. 6.46. The coating material to be sprayed is supplied to the gun with the help of powder feed unit using carrier gas (Fig. 6.46). As per requirement, quantity of the power to be supplied can be adjusted using powder feed unit. Powder particles during spraying attain high velocities (300–800 m/s) at the time impingement on the substrate.

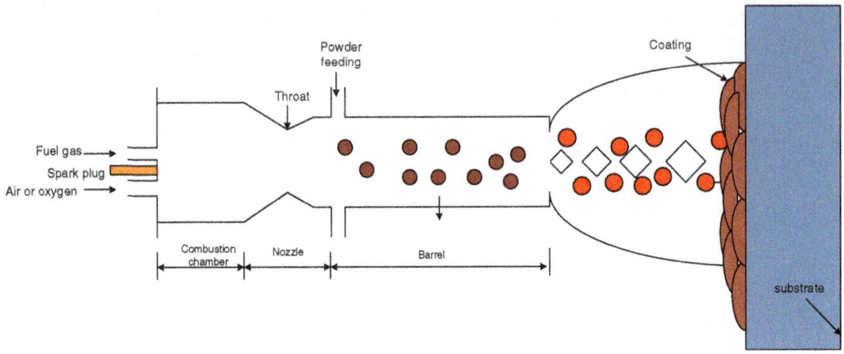

Fig. 6.46 Schematic HVOF spray system

High velocity attained by the feedstock before impinging on the substrates results in low porosity (<1%) and high bond strength (~83 MPa) of the HVOF coatings. Further, low operating temperature of HVOF (than plasma and detonation spray) reduces the thermal damage of the feedstock, for example it causes little decomposition of W_2C into WC. Besides high bond strength and low porosity, HVOF coatings possesses compressive residual stresses at the surface, which in turn improves the tensile and fatigue performance and reduces the cracking tendency.

Thus, high kinetic energy and low operating temperature of the HVOF process make them capable of producing coatings with lower porosity, higher hardness, superior bond strength, and lower decarburization than many other thermal spraying methods. The HVOF sprayed coatings have found wide application in the energy generation, cement, ship, aerospace, automobile, and general engineering industries.

Studies of high-velocity oxy-fuel (HVOF) sprayed coating of Co–Cr–W–C revealed dense and metallurgical sound coating (Sharma et al. 2009). The higher density of HVOF coating as compared to the flame sprayed coating is primarily attributed to the higher particle velocity, relatively low temperature (3000 °C), and higher impact velocity of the powder particles on to the substrate material. The coatings exhibit characteristic splat like, layered morphologies in the microstructures due to the deposition and solidification of molten or semi-molten droplets. Since the HVOF splats are less fragmented during deposition, these get flattened well and improve the contact between splats and substrate/earlier deposited material. Flattening of the HVOF splats is governed by its high velocity (*Re* (Reynolds) number). During HVOF spraying of powders, long axis of the impacted splats is oriented parallel to the substrate surface resulting in lamellar and dense microstructures (Fig. 6.47a–b). The physical features of splats dictate the porosity in microstructure of coatings which in turn affect mechanical properties and wear

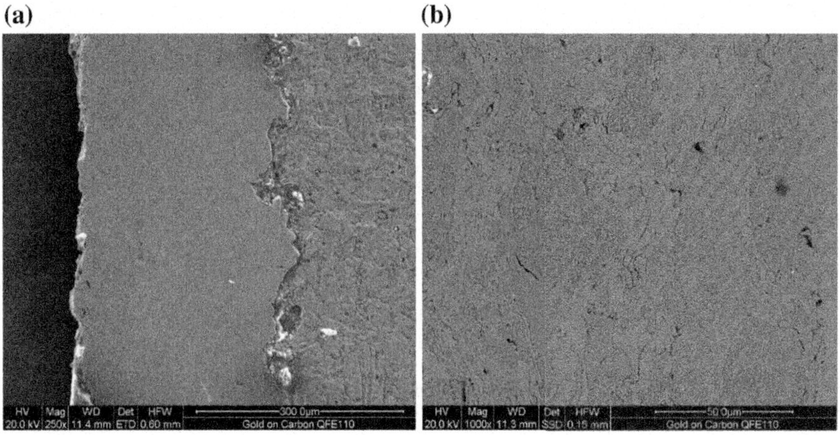

Fig. 6.47 SEM images of HVOF Co–Cr–W–C coating **a** overview of the coating and **b** closer look at middle of the coating (Sharma et al. 2009)

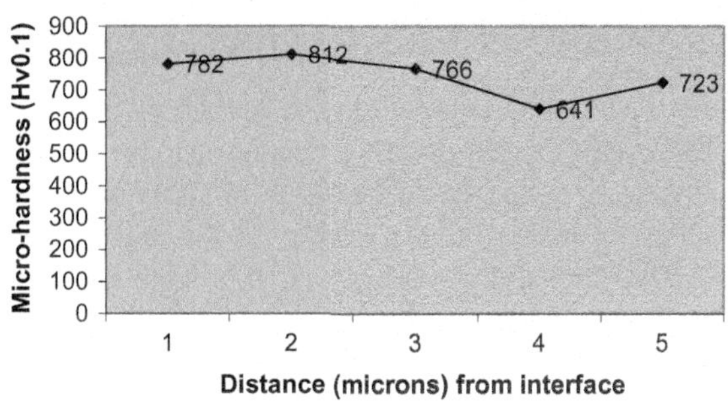

Fig. 6.48 Microhardness distribution of HVOF Co–Cr–W–C coating (Sharma et al. 2009)

resistance of coatings. The cooling rate experienced by molten Co-base alloys determines its microstructure on solidification. Higher the cooling rate, finer will be the microstructure. The substrate in the HVOF process is at a low temperature (150 °C) as compared to the molten or semi-molten splats (≈3000 °C) striking the substrate, thus leading to high heat transfer rate from splat to substrate resulting in high cooling rate of molten or semi-molten droplets.

The microhardness of Co–Cr–W–C coatings as a function of distance from coating–substrate interface and average microhardness of coatings is shown in Fig. 6.48. The overall average hardness (average of all section for each coating) of unmodified 1006 coating was found to be 745 ± 135 $HV_{0.1}$. The high hardness of the coating is attributed to solid solution hardening by W, Cr, and dispersion hardening by carbides.

6.6.3 Detonation Spray

Detonation spray process is considered to be the first high-velocity spray process, which uses a water-cooled long barrel for heating, acceleration, directing, and depositing the coating on the substrate. The heat is generated by combustion of fuel–oxygen mixture with the help of a sparks at a regular interval, and this entire process is called detonation. Nitrogen is purged after each detonation for clearing the left out "hot" powder particles and gas mixture from the chamber because these can make the process uncontrollable by an irregular detonation of the explosive mixture. Predetermined quantity of the oxygen and acetylene is fed into the tubular barrel. Simultaneously, a predetermined quantity of the coating powder is fed into the combustion chamber (Fig. 6.49).

The combustion of the oxy-fuel gas mixture generates heat, thereby resulting in the temperature rise up to 4000 °C and high-pressure shock waves traveling at high velocity (3500 m/s) along the barrel that accelerates the powder particles to be coated. The exposure of powder particles to high-temperature gases for very short period (due to high-velocity movement about 1200 m/s through the barrel) causes surface melting as well as softening of the internal region of particles. The partially melted high-velocity powder particles are directed toward the surface of substrate to be coated. The impact of high kinetic energy of the hot powder particles on the substrate results in a very dense coating with high coating–substrate bond stength. The rate of deposition of the coating is influenced by many factors such as powder particle size, carrier, composition of combustible gas mixture, frequency of detonation, distance between the barrel end and the substrate, and the gas flow rate.

The detonation frequency (1–10 detonations per second) is determined by the coating thickness to be developed and the hardness of coating material. The extremely high kinetic energy of the partially melted particle (plastic state) after the impact results in the coating of good adhesive strength, low porosity, and compressive residual stresses at the surface.

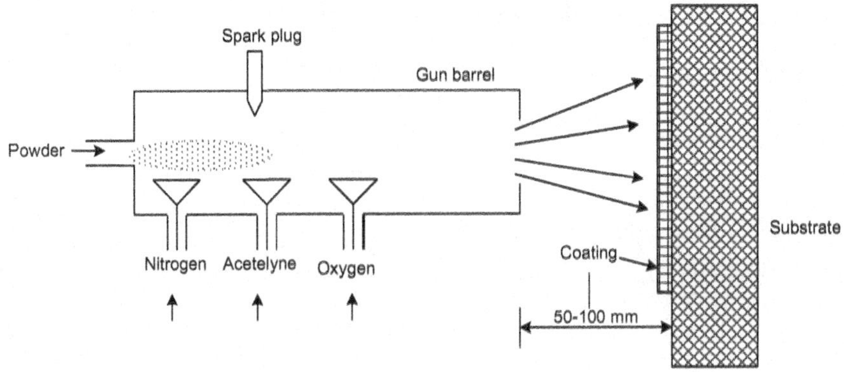

Fig. 6.49 Schematic of detonation gun spraying process

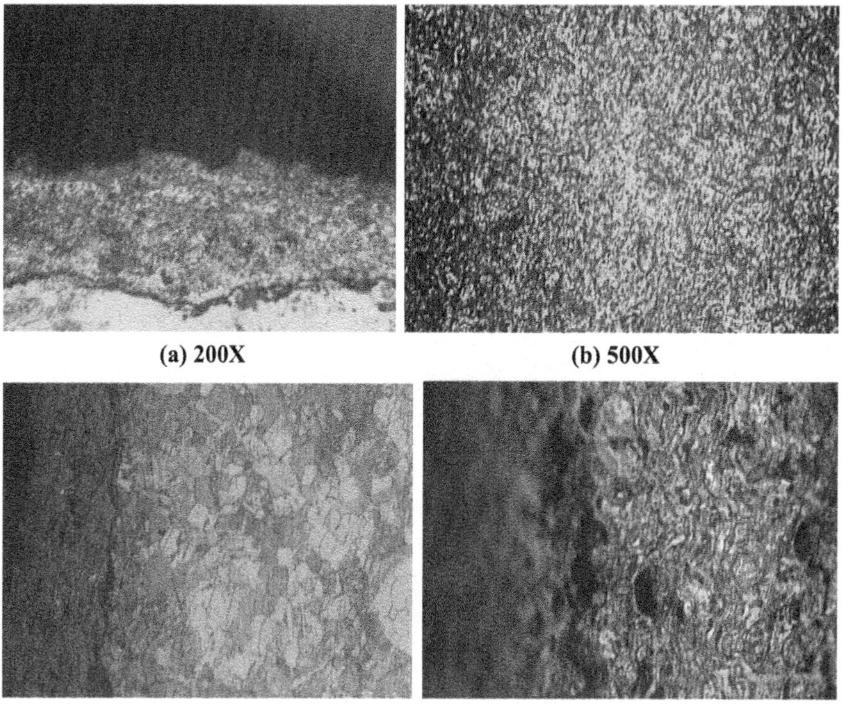

(a) 200X (b) 500X (c) 200X (d) 500X

Fig. 6.50 Microstructure of **a–b** WC–CO–Cr and **c–d** Ni–Cr coating deposited by detonation gun spraying (Arji 2006)

The detonation equipment having a horizontal orientation consists of a double-walled barrel (length 1800 mm and inner diameter 22 mm), a combustion chamber and powder feeder, apart from control panels to regulate gas flows. Arji and Dwivedi (2006) studied microstructure, hardness and erosion behavior of Ni–Cr and WC–Co–Cr coating deposited on austenitic stainless steel substrate by detonation spraying. Typical parameters used for depositing these coating are powder particle size of 35–45 μm, spray spot diameter 25 mm, spraying distance of 200 mm, coating thickness per shot of 5–6 μm, shots per second of 4, total coating thickness of 300 μm, and 120 μm for Ni–Cr and WC–Co–Cr, respectively. Micrographs of the detonation sprayed coating showed splats and defect-free sound deposition (Fig. 6.50a–d).

Microhardness of WC–Co–Cr and Ni–Cr coating was 1166 HV_{100} and 560 HV_{100}, respectively. It has been found that detonation gun coating increases the abrasive wear resistance of austenitic stainless steel significantly. Coriolis erosion wear test at 800 rpm caused wear scar of different areas on austenitic stainless steel (ASS) substrate, Ni–Cr and WC–Co–Cr coating (Fig. 6.51). The wear scar areas decrease with increase of radial distance of the specimen from the

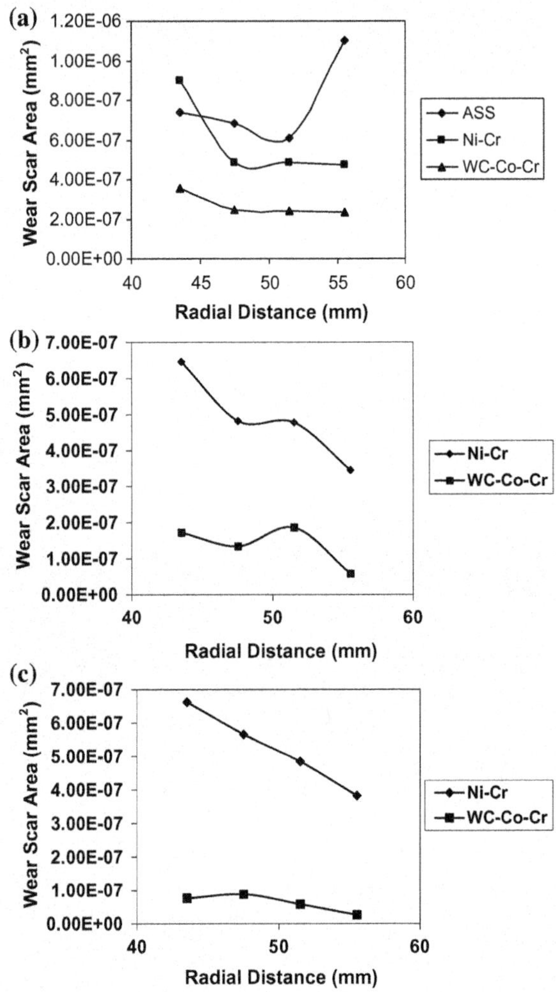

Fig. 6.51 Wear scar cross-sectional area versus distance from rotor center at a speed of **a** 800 rpm, **b** 1500 rpm, and **c** 2000 rpm (Arji 2006)

center during erosionv test. Distribution of erosive wear features on austenitic stainless steel surface is uneven. The WC–Co–Cr coatings subjected to erosion fewer wear scar areas than Ni–Cr coating and ASS substrate. The wear scar areas of WC–Co–Cr and Ni–Cr reduce with increase in radial distance during erosion test at all speeds (Fig. 6.51a–c).

The SEM images of WC–CO–Cr showed internal surface deformed zones and craters in the brittle phase (Fig. 6.52a–b). The SEM micrographs of eroded surface of Ni–Cr coating showed plastically displaced and plowed sideways to form impact crater side ridges (Fig. 6.53a–b).

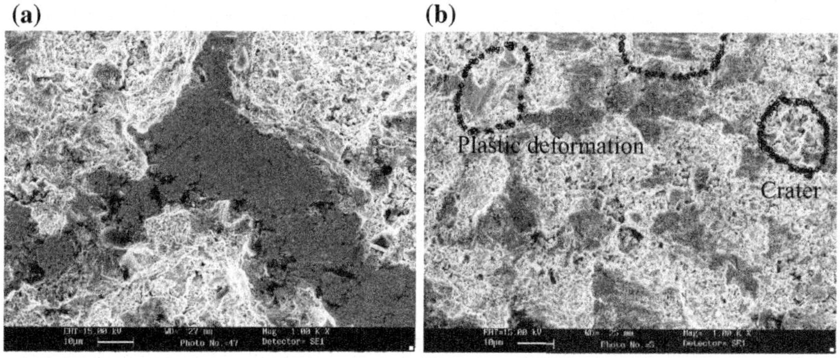

Fig. 6.52 SEM images of **a** as-sprayed-coated surface and **b** eroded surface of WC–Co–Cr coating (Arji 2006)

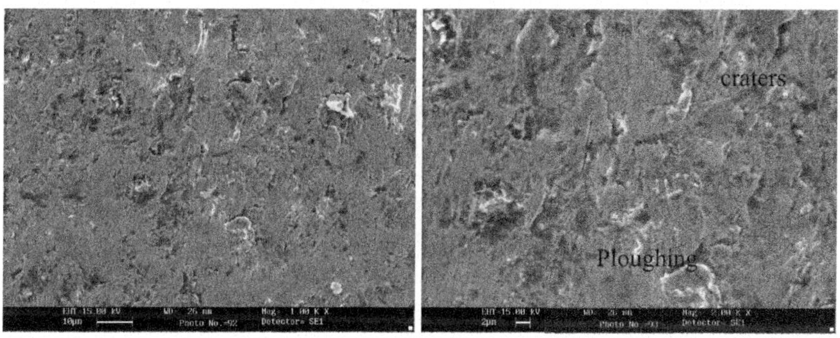

Fig. 6.53 SEM Micrograph of eroded surface of Ni–Cr coating on stainless steel substrate (Arji 2006)

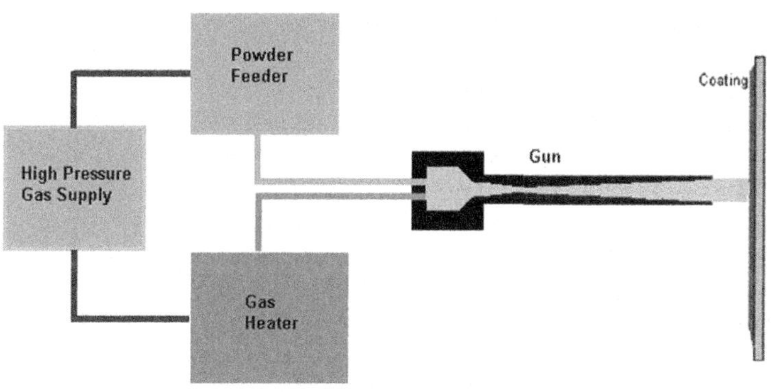

Fig. 6.54 Schematic diagram of cold spray system

6.6.4 Cold Spraying

Cold spraying is a result of development in the direction of high kinetic energy coating processes like HVOF. It mostly relies on the high velocity (500–1500 m/s) imparted to the slightly heated powder particle using compressed gas (He). The powder particles are heated using hot gases (300–400 °C), which are fed into the gun (Fig. 6.54). The mixture of compressed gas, heating gas, and powder particles passes through a specially designed convergent–divergent nozzle for accelerating the power particles and impacting onto the surface of the substrate at high velocity. The powder particles impacting a substrate deform plastically and develop bond with the substrate to form a coating. An optimum combination of particle size and density, velocity and temperature plays an important part in achieving the desired coating characteristics.

Powder particles retain their characteristics, as they remain in solid state. The absence of melting and heating to a very low temperature avoids any thermal damage to coating material in the form of oxidation and undesirable metallurgical charges. The impact of high kinetic particles in the solid state produces plastic strain in the splat; therefore, the cold sprayed coating tends to develop compressive residual stress at the surface. Further, the impact of particle (having low temperature) with substrate at high velocity breaks the layer of impurities and films present in the form of oxides at the surface of powder particles which in turn promotes a bonding among the clean splat surfaces. In this process, the bonding between the powder particles and the substrate initially becomes mechanically interlocking. Owing to this nature of the process where the bond is developed by plastic deformation of powder particles, only soft and ductile metal like aluminum, copper, titanium, stainless steel, and their alloys can be sprayed using cold spray process. Further, the substrate must be strong and rigid enough to withstand the powerful action of high-velocity jet of cold spray process else it can erode or distort during the spraying.

Advantages

- No thermal, metallurgical (composition/phases) damage to coating materials;
- Low-temperature process
- No melting of power particle
- Very little or no oxidation of powder particles
- Absence of solidification stresses;
- Introduces the favorable compressive stress;
- Lowering the requirement of surface preparation of substrate using grit blasting.

Limitation

- Process is limited to soft and ductile metal as hard and brittle materials can be applied only as composite materials with soft and ductile matrix.
- Substrate should also show some plastic behaviors; hence, all substrate materials cannot be coated with same ease.
- High gas consumption as He is very expensive.
- Limited data is available on the properties of coatings.

Application

- To develop coatings that are expected to perform well in the absence of oxidation such as corrosion protection, enhanced electrical, and thermal properties.
- To use preplacement materials as solders and coating for better results.

6.6.5 Electric Arc Wire Spray Process

Electric arc wire spray process works on the simple principle of using heat produced by an electric arc established between two electrode wires made of material to be coated. The molten metal produced from melting of electrode wires is atomized and accelerated toward the substrate using compressed air. The impact of the high-velocity particles with substrate forms splats, and these on solidification results in coating. Arc is developed only between two electrodes, and the substrate does not form part of any electric circuit. For this reason, the standoff distance between spray torch and substrate allows flexibility in the movement. The application of DC results in higher melting of anode than cathode; therefore, AC is preferred for the development of coating using this process. High temperature facilitates the spraying of wide range of electrical conducting material only. This process is primarily used for developing corrosion-resistant coating such as zinc, magnesium. The bond strength, deposition rate, and porosity of electric sprayed coatings are found much higher than the respective flame spray coatings.

6.6.6 Plasma Arc Spray

The principle of plasma spray process is similar to that of PAW. In this process, the heat of the plasma is produced by passing plasma forming gas such as Ar, N_2, air, through an arc established between tungsten electrode and water cooled nozzle. Energy density of plasma is increased by forcing the plasma to pass through constricting nozzle, which in turn not only increases the temperature but also the velocity of plasma gases. Feedstock in the form of wire or power is fed into the plasma region. High temperature associated with plasma facilitates the melting, softening, and spraying of even high-temperature metals such as tungsten, nickel, cobalt, and ceramic material like zirconia.

Mostly, the plasma spraying is performed under normal ambient condition and is called as atmospheric plasma spray (APS). However, to develop the quality coating using plasma spray process accordingly, the molten particles are protected from contamination by atmospheric gases using vacuum chambers. The plasma spray performed under vacuum condition is called vacuum plasma spray (VPS) and if the chamber is backfilled with low-pressure gas; then, accordingly the process is termed as low-pressure plasma spray process. High temperature and reasonably high particle velocity during plasma help to develop coating having high strength, low porosity, and high coating–substrate bond strength. These properties become comparable to any of the combustion-based thermal spray processes except HVOF and detonation spray. Both electric arc wire spray and plasma spray process keep the substrate at a very low temperature, which in turn reduces the metallurgical damage of the substrate, residual stress, and distortion tendency.

6.7 Electroplating

Electroplating is opposite to the galvanic cell and is used to develop coatings of electrically conducting metals like Au, Ag, Cu Zn, Ti, Cr for variety of purposes such as increasing shining, wear and corrosion resistance, electrical conductivity, and buildup dimensions. For electroplating, galvanic cell is developed by using electrical conducting substrate (cathode) and coating material (anode). Furthermore, suitable electrolyte for the flow of current and power source to establish the potential difference for the flow of current are used (Fig. 6.55). Once the electrical circuit is completed, the dissolution of metal starts at the anode and begins to get deposited on the cathode as per current density between the anode and cathode. The projections on substrate lower the gap between anode and cathode, which in turn increases the current density; therefore, the metal is deposited at higher rates at the projected areas in comparison to other areas.

The substrate and coating metal are immersed in electrolyte solution having dissolved metal salt and other ions. These ions facilitate the flow of electricity through the electric circuit. DC is fed to the coating material, which oxidizes metal atoms and dissolves them into the electrolyte, whereas the dissolved metal ions are reduced at the cathode and facilitate the deposition of metal onto the substrate. During electroplating, the rate of anodic dissolution is found equal to the rate of deposition of metal onto the substrate.

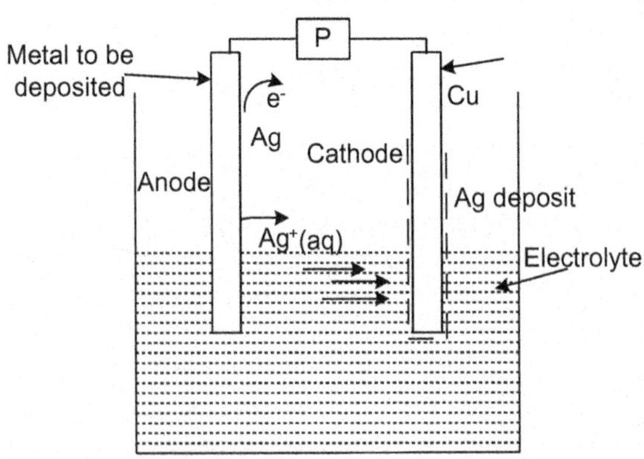

Fig. 6.55 Schematic diagram showing the principle of electroplating process

The electroplating of silver involves the application of silver that is used as the anode, and the electrolyte solution contains the ion of the silver. Silver goes into solution by dissolution of the anode, and the same is deposited onto the cathode. The concentration of silver ions in the electrolyte is maintained. However, this process can be used for coating of electrical conducting metals only. Typical reactions of iron, aluminum, and silver plating are shown below.

$$Ag^{-} \rightarrow Ag^{+} + e^{-} \text{ (anode)}$$
$$Ag^{+} + e^{-} \rightarrow Ag(s) \text{ (deposited on cathode)}$$
$$Eg\,.\,Fe^{2+}(aq.) + 2^{e^{-}} \rightarrow Fe(s)$$
$$Al^{3+} + 3^{e^{-}} \rightarrow Al(s)$$

6.8 Electroless Process

The electroless process is basically an autocatalytic process that primarily involves chemical reactions for developing coatings of Ni and Ti metals without any requirement of power source and electrolyte. It uses an aqueous solution composed of the metal to be deposited along with hypophosphite, which acts as a reducing agent during the chemical reactions (Fig. 6.56). The aqueous solution is reinforced with hard particles like WC, TiB_2 for developing coating of composite material that constitutes Ni–WC, Ni–TiB_2. The substrate to be coated is dipped into the solution, and by autocatalytic reaction, coating is deposited. The deposition rate of process using low-temperature solution (30 °C) is very less (~1 µm/h). Increase in the solution temperature (to about 90 °C) increases the deposition rates (~10–20 µm/h) appreciably. Electroless coating can be applied on both the metallic and non-metallic components. This process is used for developing the coatings onto the substrate, where thermal spray cannot be applied due to the possibility of thermal or mechanical damage. However, the coating material for electroless process should be electrical conducting.

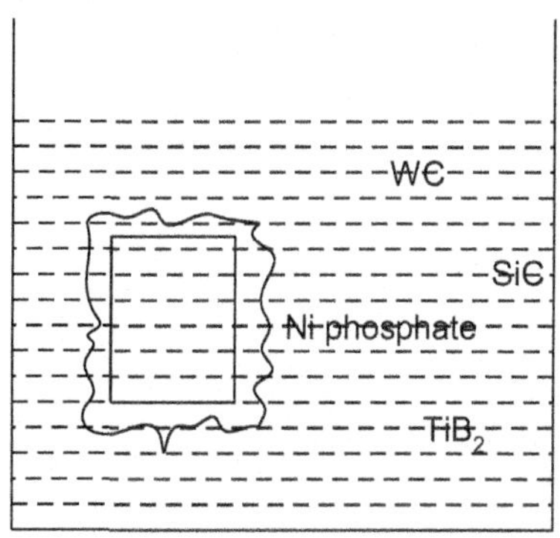

Fig. 6.56 Schematic diagram showing principle of electroless process

Common hypophosphite used—Na, PO_2, H_2O

$$2H_2PO_2^- + 2H_2O + Ni_2^+ \rightarrow Ni + H_2 + 2HPO_3^- + 2H^+$$

Questions for self-assessment

1. What is fundamental approach of surface modification by developing coating and cladding?
2. List the methods used for surface engineering by developing coating and respective range of thickness of layer achieved.
3. Explain the factors affecting characteristics and soundness of the coating and cladding developed for surface modification.
4. Explain the principle of hardening the surface of steel by carburizing.
5. What is need to (a) protect molten metal and (b) control the dilution during weld surfacing?
6. How does the welding process used for surfacing affect the quality of engineered surfaces?
7. What is role of heat input and number of layers on characteristic of the weld surfacing?
8. How do the welding parameters affect the geometry of weld surfacing developed by SAW?
9. Explain the principle of pulse GMAW using schematic diagram for weld surfacing.
10. How can pulse GMAW process be different from conventional GMAW weld surfacing?
11. How can metal-cored arc welding provide more flexibility for developing weld surfacing of varying composition?
12. Why do carburizing and nitriding improve the fatigue life?
13. How can GTAW cladding be beneficial over SMAW cladding?
14. Explain the principle of PTAW for surface modification using suitable schematic diagram.
15. What is laser cladding? Give advantage of Laser cladding over conventional cladding method.
16. Describe the principle of thermal spraying process?
17. Explain the factors affecting the quality, and soundness of thermal spray coating.
18. Explain the principle of following processes using suitable schematic diagram.
19. Flame spraying.
20. High-velocity oxy-fuel spraying.
21. Detonation gun spraying.
22. Explain the approaches which can be used improve the characteristics of the flame sprayed coating.
23. How does ceria affect characteristics of flame sprayed coating of Ni base alloy?

24. How does chromium carbide addition affect the abrasive wear behavior of flame sprayed coating of Ni–Cr–B–Si alloy?
25. Describe principle of cold spraying process and its advantages over other processes like flame spraying.
26. Describe principle of electroplate and electroless plating process using suitable schematic diagrams.

Chapter 7
Characterization of Engineered Surfaces

7.1 Introduction

The performance of the engineered surfaces is determined by various chemical, mechanical, metallurgical, tribological, dimensional, and surface properties. Among these properties, probably the first four types of properties, namely chemical, metallurgical, mechanical, and tribological are interrelated sequentially. However, the relative importance of the above properties on the performance of the modified surfaces varies with the application and service related requirements appreciably. On the basis for these properties, decision is taken whether engineered surfaces will be fit for the target application or not, to assess the appropriateness of an engineered surface for the target application. The characterization of the engineered surfaces is performed to achieve the following: (a) assess the suitability of the modified surfaces for a particular application and (b) quantify the properties of modified surfaces. In the present chapter, various methods of characterizing the common properties of engineered surfaces have been described.

7.2 Characterization of Surface Properties

The surface characteristics such as roughness, hardness play an important role in determining the tribological performance of modified surfaces because these properties decide the ease of nucleation of crack and indentation to cause abrasive, surface fatigue, and erosive wear.

D. K. Dwivedi, *Surface Engineering*,
https://doi.org/10.1007/978-81-322-3779-2_7

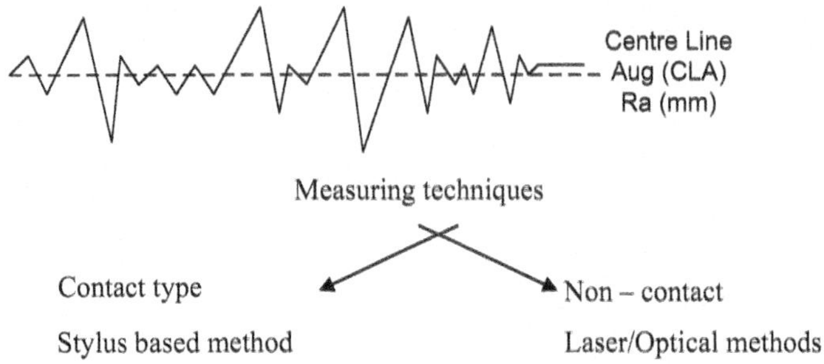

Fig. 7.1 Schematic of surface irregularities and measuring techniques

7.2.1 Surface Roughness

The surface roughness of a component is an important characteristic as it determines (a) wear and friction behavior of coatings and modified surfaces, (b) resistance to stress corrosion cracking (greater surface roughness more will be the sites for nucleation of cracks), and (c) fatigue life. The measurement of surface roughness means the quantification of magnitude of ups and downs present in the form of peaks and valley of asperities on the surfaces (Fig. 7.1). Center line average value is one of the most commonly used measures of surface roughness that provides the average of height of all peaks and depth of all valleys present over a given sampling length with respect to a center line of surface variation (an imaginary line). There are two broad categories of approaches used for surface roughness measurement based on the fact whether the surface to be tested is in physical contact during the test or not (a) contact type and (b) non-contact type.

7.2.1.1 Stylus-Based Approach

Many surface roughness measurement systems have been developed using the stylus-based approach of surface roughness quantification. In this approach, a sharp-tipped stylus navigates across the surface to be assessed for surface roughness. As per the asperities (resulting peaks and valleys) present on the surface, stylus moves up and down while traversing on to the surface. This up and down movement of stylus is used to quantify the roughness (Fig. 7.2). Since magnitude of

Fig. 7.2 Schematic of stylus-based approach

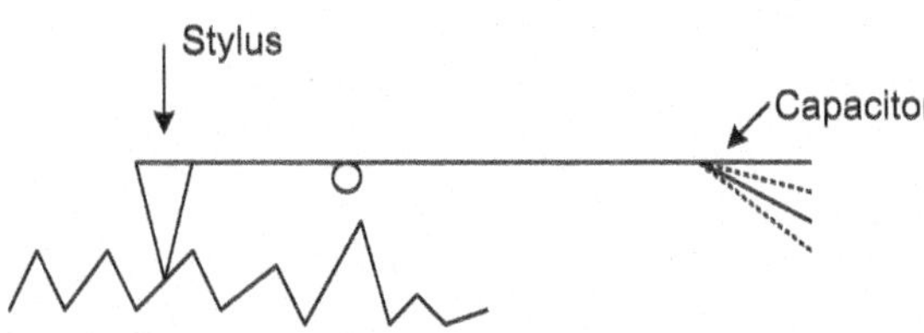

the movement of stylus is extremely small, it is increased using several approaches such as lever principle, electrical signals using suitable capacitors. This approach is suitable for measuring the surface roughness belonging to the order of microns. The stylus-based approach is preferred because it is highly accurate and gives a direct measurement through the up and down movement of stylus. For measuring the surface roughness of extremely thin (of few nm) films and coatings, atomic force microscope (AFM) is used which can measure roughness of nanolevels. However, the stylus-based approach suffers from various limitations such as the following: (a) time consuming owing to slow speed of stylus movement, (b) short sampling length (50–800 μm) as only a small area is considered while scanning the surface during the measurement; hence, surface roughness test needs to be repeated many times for large surface area objects, and (c) inability to measure the surface roughness of moving objects like during rolling or extrusion processes.

7.2.1.2 Laser-Based Approach

Laser-based approach does not involve any physical contact between the measuring system and surface of the object. In this approach, laser beam is directed on the surface from a suitable laser source, wherefrom laser rays are reflected according to the ups and down present on the surface (Fig. 7.3). The reflection pattern of laser rays is recorded, observed, and related to the extent of roughness of the surfaces. However, this method is not very accurate, as the relationship between reflection pattern and roughness is not direct. This method is very effective in measuring the surface roughness of object, where physical contact with surface of the object is not practicable.

7.2.2 *Confocal Scanning Microscope*

The normal laser-base technique gives 2D reflection patterns. The confocal scanning microscope method performs sequential depth-wise scanning of the surface

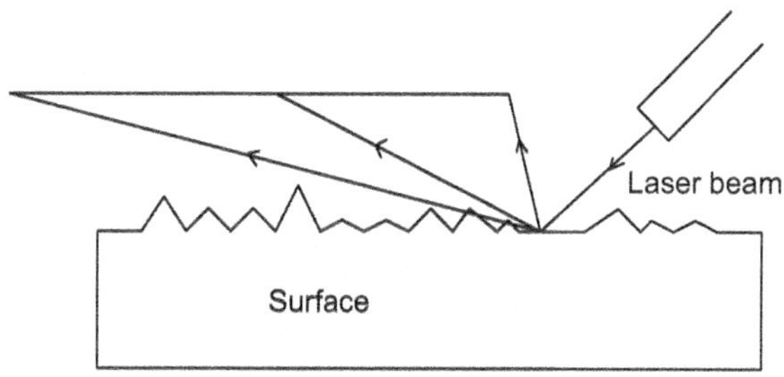

Fig. 7.3 Schematic of laser-based approach

and provides 3D reflection pattern at different surface depths. This technique is widely used to measure the roughness in remote places and areas of component which are unreachable.

7.3 Thickness of Coatings and Films

Friction force generated during interactions between the mating components results in shear stresses at the surface and the subsurface region. Thickness of coatings or modified zone should be such that the maximum shear zone falls within the coating. The occurrence of maximum shear zone at the coating–substrate interface leads to ease failure of coating or films by delaminating and spalling. Further, indentation due to hardness of particles and objects during the service should not reach up to the substrate, as it increases failure tendency of coatings and so reduces tribological life of functional surfaces. These requirements make it necessary to measure the thickness of modified or engineered surface. The following three methods are commonly used for measuring the thickness of the coating: (a) weight measurement of the coating deposited, (b) surface indentation method, and (c) ultrasonic method. Each method has its own merits and limitations.

7.3.1 Weight Measurement of Coating Deposited

This method is based on the simple approach of measuring the coating deposited on the component before and after the surface modification. Subsequently, thickness (t) is obtained from weight (W), density (ρ), and dimensional features (length "l" and width "b") of coating using the equation given below. This method is found to be accurate above 100 nm and is determined by the least count of weighing balance.

$$\Delta W/\rho = V = e \times b \times t$$

7.3.2 Indentation Method

In this method, an indentation is made onto the modified and coated surface using a large diameter (D) ball indentor (20–30 mm), and suitable load is selected in such a way that indentation is made onto the substrate after penetration of the coating or films (Fig. 7.4). The diameter of indentation at the top surface of the coating and that at the coating–substrate interface (d) is measured. Thickness (T) of the coating

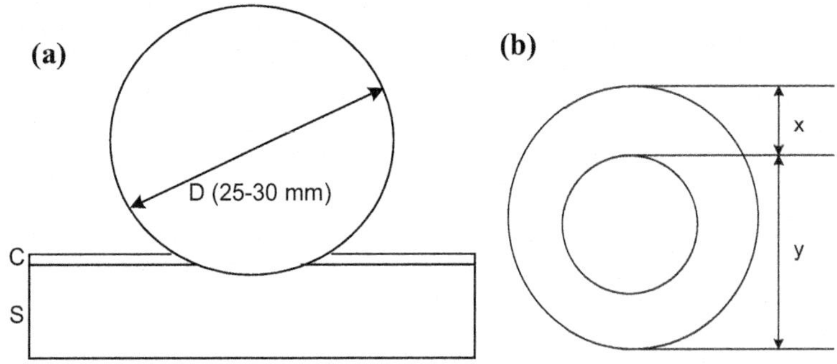

Fig. 7.4 Schematic of principle of indentation technique **a** front view and **b** top view of indentation

is calculated using the equation ($T{:}xy/D$), where the parameters x and y are obtained from the indentation measurement correspond to the diamter of indentation at the surface and coating-substrate interface respectively. This method however, is not accurate for measuring coating thickness greater than 100 μm.

7.3.3 Ultrasonic Testing

This method is very efficient and easy to apply. It is based on the simple principle of transmission of ultrasonic vibrations into the coated samples to be tested and subsequenty their reflection from the interfaces. These vibrations are reflected as soon as these come across the change of mediums such as air–metal at the top surface, coated–substrate at interface, substrate–air at bottom surface and through defects in coating in the form of inclusions, porosity, cracks, etc., if any. The first reflection of vibration from air–metal interface at the top surface is observed as a strong and high peak in oscilloscope. The reflection of vibrations from coating–substrate interface in the case of coated specimens if there is no internal defect is observed as the second peak (somewhat lower/weaker than first peak) in oscilloscope, and the reflection of vibrations from substrate–air interface at bottom surface is observed as the third and strong peak (like first peak) (Fig. 7.5a, b). A properly calibrated ultrasonic testing system directly gives thickness of the coating from the relative spacing between first and second peak, whereas the gap between second and third peak indicates the thickness of substrate. It is required to smoothen and finish the top surface of coated specimen, so as to obtain good physical contact between ultrasonic vibration probe and surface of the coating to be tested. A thin layer of liquid (water or oil) called coupler is generally applied between the probe and surface of coating to reduce the losses of ultrasonic vibrations and to make more effective transmission of vibrations into the coating–substrate system to be tested.

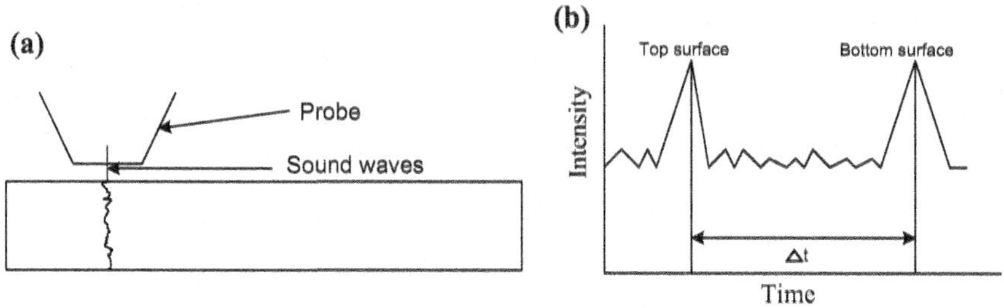

Fig. 7.5 Schematic of ultrasonic technique **a** ultrasonic vibration directed on to the component and **b** intensity signals reflected and received back as function of time

7.4 Bond Strength of Coating–Substrate

The surface modification of substrate using films and coating is performed by different processes such as weld surfacing, thermal spraying, chemical and physical various deposition techniques. These techniques of surface engineering are based on different principles of bonding between substrates and coating or films such as mechanical interlocking, diffusion, and metallurgical bonding. Mechanical interlocking contributes appreciably in cold spray and high-velocity oxy-fuel spray processes for developing bond, while weld surfacing and spray and fuse processes involve metallurgical bonding at coating–substrate interface. Mechanical interlocking offers comparatively weaker bondage between the coating and substrate than weld surfacing and other diffusion-based processes. In some of the diffusion-based processes (like CVD and PVD), the modified surface becomes an integral part of component, so there is no issue of bond strength, but in case of thermal spray coating, bond strength must be assessed. The poor bond strength can lead to easy spalling and chipping off coatings from the substrate during service under the influence of tangential force acting during sliding, abrasion, erosion, etc., and stress induced due to differential expansion and contraction owing to heating and cooling during service (Fig. 7.6). Coatings are expected to have minimum acceptable bond strength of the order of 70 MPa. Bond strength of the coating is evaluated using three common methods: (a) bend test, (b) scratch test and (c) tensile test.

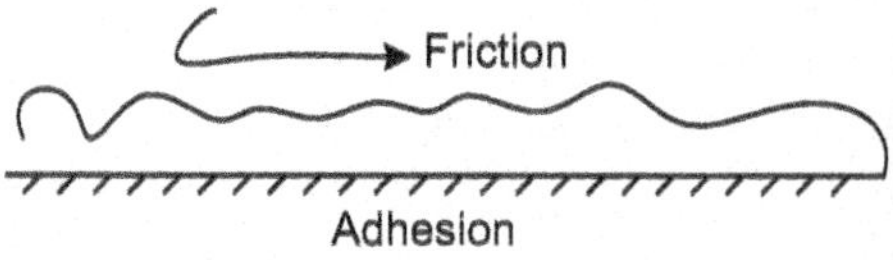

Fig. 7.6 Schematic showing the need of bond strength owing to friction

7.4.1 Bend Test

In bend test, coating–substrate system (coated sample) is bent over a form block under controlled conditions and initiation of crack at the coating–substrate interface is observed carefully during bending (Fig. 7.7). The angle at which cracking or delamination or chipping off is triggered during bending is used as a measure of bond strength. Low angle of bend at which chipping off or cracking occurs at the interface indicates poor bond strength. A bend of 180° shows perfect bonding between coating and substrate.

7.4.2 Scratch Test

Scratch test basically determines the ability of the coating to get chipped off under shear load conditions. The shear stress on the coating is applied through an indenter. Indenter is given two movements simultaneously; one is longitudinal along the surface of coating, and the other is vertically downward (Fig. 7.8). This kind of movement of indenter develops scratch of increasing depth. The force required (recorded continuously) to develop such kind of scratch increases almost linearly with the increase in length of scratch until the commencement of delamination or spalling of coating from the substrate. Deviation in force versus scratch length curve is noticed as along with the spalling or chipping off coating. The maximum force that a coating sustains (without deviation in curve) during the scratch test is used as a measure of bond strength of coating with substrate (Fig. 7.9). Higher the force required to start the delamation, greater is the bond strength. With further increase in indentation, force decreases abruptly on approaching the coating–substrate interface, as the substrate needs lower force than the coating.

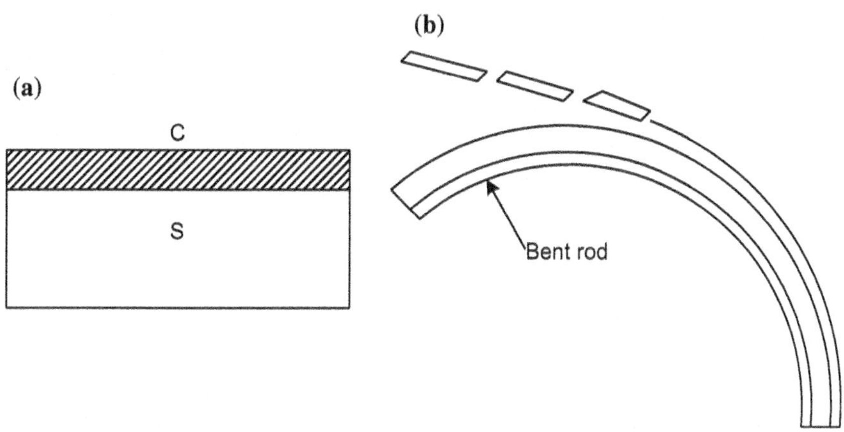

Fig. 7.7 Schematic of bend test approach **a** coating on the substrate and **b** coating delaminated from the substrate

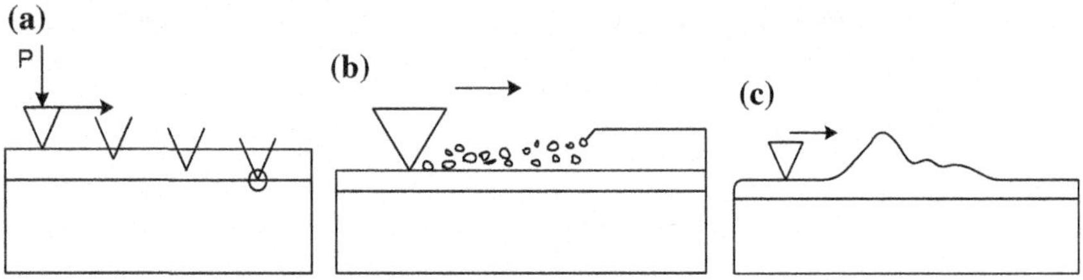

Fig. 7.8 Scratches testing **a** scheme of movement of indenter, **b** chipping in case of brittle coating, and **c** deformation in case of ductile coating

Fig. 7.9 Schematic diagram showing the force versus length of scratch relationship

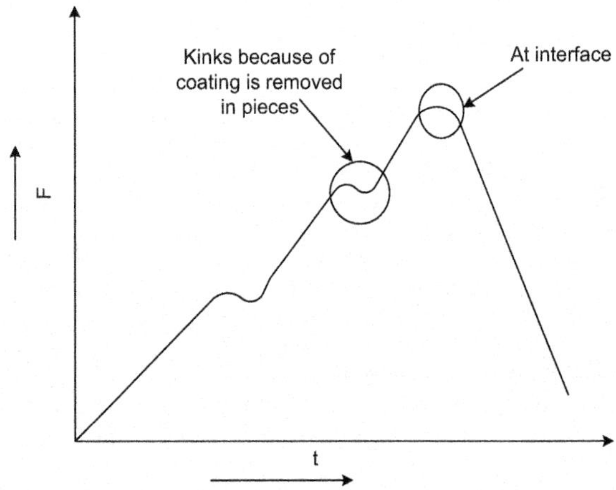

7.4.3 *Tensile Test*

Tensile test method for assessing the bond strength of the coating with substrate is similar to the tensile test used for determining the tensile properties of materials. The application of external tensile load results in failure from weak region, and this approach is indirectly used to establish the bond strength of coating–substrate interface. To evaluate the bond strength of coating with substrate, an adhesive joint is developed between a) the surface of coating on the substrate and b) a structural member made of substrate (Fig. 7.10). The adhesive having strength greater than 70 MPa is used for developing such joint. Through application of the tensile load, the joint failure location is identified during tensile test. Failure from the coating–substrate interface suggests that the bond strength of coating–substrate interface is lower than 70 MPa. However, failure from adhesive joint between coating surface and another structural member indicates that the coating–substrate bond strength is greater than 70 MPa and the same is acceptable as per international practices.

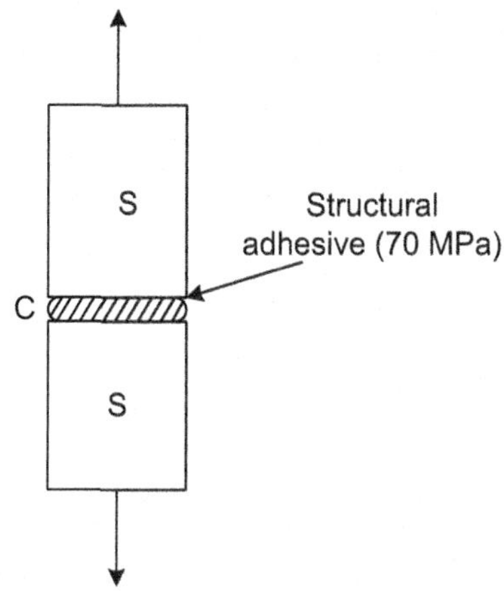

Fig. 7.10 Schematic of tensile test approach

7.5 Soundness of Modified Surfaces Using Non-destructive Testing (NDT)

To determine the presence of surface and sub-surface imperfections, non-destructive testing (NDT) can be carried out using different techniques as per needs. Apart from the visual inspection, many NDT methods including dye penetrant test (DPT), magnetic particle test (MPT), eddy current test (ECT), ultrasonic test (UT), radiographic test (RT), etc., are utilized in manufacturing industry for assessing the soundness of engineered surfaces. In the following section, the principle and capability of some NDT methods have been described.

7.5.1 Dye Penetrant Test

Dye penetrant test (DPT) is one of the simplest NDT methods that are primarily used for detecting the presence of surface defects only. In this method, surface to be tested is applied with a thin liquid of low viscosity and low surface tension containing suitable dye (Fig. 7.11). The thin liquid penetrates the surface (by capillary action) into fine cavities, pores, and cracks if any present on the modified surface. Excess liquid present at the surface is wiped out. Following this, suitable developer like talc or chalk powder is sprinkled over the surface. Developer sucks the thin liquid with dye, wherever it is present inside the defects on the surface. Dye with liquid changes color of developer and thus indicates location and size of surface defects.

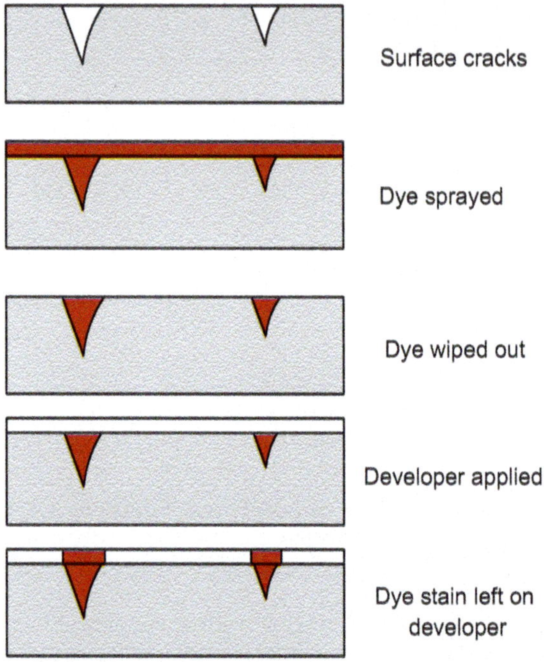

Fig. 7.11 Schematic of stepwise approach of dye penetrant test

7.5.2 Magnetic Particle Testing

Magnetic particle testing (MPT) is mainly used for assessing the surface and near-surface defects in a magnetic material. It is based on the simple principle of the flow of magnetic line of forces. Magnetic flux flows easily through metal from South Pole to North Pole. The component to be evaluated is magnetized using electrical energy or suitable permanent magnet. The electromagnetization is performed using suitable yoke which is applied across the location to be tested. The presence of any discontinuity in the form of crack, porosity, near-surface defects in the path of flow of these lines results in the leakage of magnetic flux and forms two additional poles. The magnetic powder particles (in dry form or suspension form in thin liquid) are sprinkled over the surface of magnetized components to be tested. The magnetic particles tend to migrate toward the location wherever there is leakage in the magnetic flux and then get piled up. The particles align along the discontinuities at the surface or near the subsurface discontinuities. The location and pattern of piled up magnetic powder particles suggest the location, size, and type of discontinuity present on the surface or near surface region. Hazy pile of powder particle indicates the presence of subsurface defect. The formation of very thin line of powder particles suggests the presence of crack with details of size and location of defect. However, this method of testing is fit for ferromagnetic metal only (Fig. 7.12).

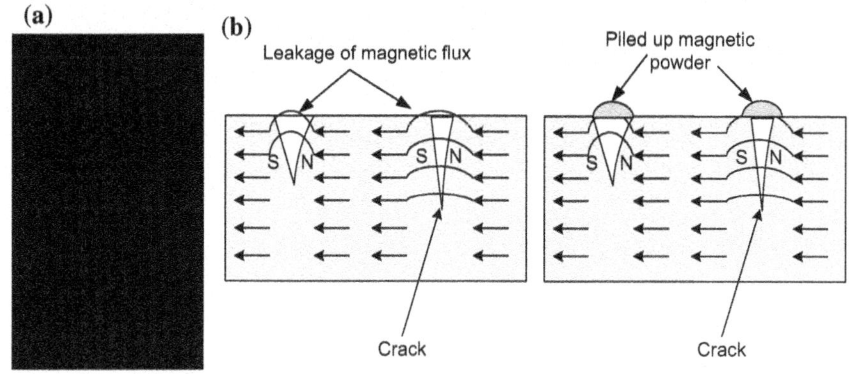

Fig. 7.12 Schematic of magnetic particle test approach **a** leakage of magnetic flux before sprinkling of ferromagnetic power and **b** piling up of ferromagnetic powder

7.5.3 Ultrasonic Testing

Ultrasonic testing (UT) is one of the most popular, quick, cost effective, and capable methods of NDT, as it not only indicates the presence of discontinuities but also suggests their location using ultrasonic vibrations. Ultrasonic vibrations have the capability to penetrate the metals and are reflected as soon as they come across a change of medium, for example, metal to air, air to metal, metal to discontinuities etc. This reflection characteristic of ultrasonic vibrations from the interfaces of change in medium is mainly exploited for detecting the presence or absence of discontinuities. The application of ultrasonic vibrations in a sound metal system at the sources produces two interfaces; one is at the top surface due to change of medium from air to metal, and the other is at the bottom surface due to change of medium from metal to air (Fig. 7.5). For evaluating the modified surfaces in consideration, the ultrasonic vibrations are used in two ways: (a) transmission and (b) reflection. All these methods are very effective for parallel surface components.

7.5.3.1 Transmission Approach

The transmission approach of ultrasonic testing uses two separate devices, namely transmitter of vibration and receiver. Transmitting probe generates and sends the ultrasonic vibrations, and receiver gets these vibrations to the other end. Therefore, transmission approach needs access to both the sides of the components to be tested. Inputs from transmitting and receiving probes are given to oscilloscope. Metal system without discontinuities shows the two peaks in oscilloscope, that is, one from the top surface and another from the bottom surface. In the presence of discontinuity in the metal being tested, ultrasonic vibrations are reflected so they do not reach up to the receiving end, and so no signal is received. Under this condition, only one peak in observed in the oscilloscope and absence of another peak from bottom surface suggests the presence of discontinuity in the metal being tested. One by one the entire surface area of the component to be tested is scanned using transmitting and receiving probes. However, transmission approach is not very useful for two reasons: (a) the requirement of access to both sides of component to be tested and (b) the difficulty in placement of receiving probe in line of ultrasonic vibrations, especially in case of components having thick sections.

7.5.3.2 Reflection Approach

The reflection approach uses single probe which plays the role of transmitter as well as receiver of ultrasonic vibrations. In a metal system without discontinuities, the application of ultrasonic vibrations shows two peaks in oscilloscope, one from the top surface and another from the bottom surface like the transmission approach. In the presence of discontinuity in the metal being tested, ultrasonic vibrations are reflected by metal-discontinuity interface. Vibrations reflected from the discontinuity shows additional peaks between the top and bottom surface peaks in the oscilloscope. Relative location of the intermediate peaks (between the top and bottom surface peaks) suggests the distance of discontinuity from the surfaces. The reflection approach overcomes both the limitations of transmission approach, as it uses single probe, so it does not require (a) access to both sides of the component to be tested and (b) alignment of transmission and receiving probes.

7.5.3.3 Pitch-Catch Method

In pitch-catch method, ultrasonic vibrations are transmitted from 45° or 60° probe to the surface of the material to be tested. Reflected vibrations from the other reflecting

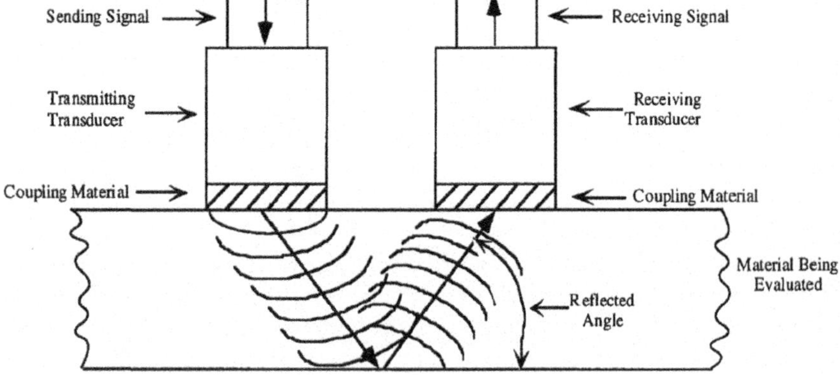

Fig. 7.13 Schematic of ultrasonic test approach for discontinuities

surface or discontinuity are used to evaluate the weld joints and other parallel-sided surfaces (Fig. 7.13).

Coupler

For the effective transmission of ultrasonic vibrations from the source probe to the metal, generally a fluid mostly in the form of water or low-viscosity liquid called coupler is used. The coupler ensures proper contact and transmission of vibration from source probe to metal with minimum losses. Water is considered as the best coupling media to use with most construction materials because of its easy availability, low viscosity, and relatively safe nature. In the pitch-catch method, a water-based gel has proven to be the most practical coupling agent.

Precautions

Ultrasonic testing needs precautions for increased effectiveness, which are

- Use of the suitable coupling media for matching of acoustical impedance of the sensors with the substrate/coating.
- Use of air–coupling for moisture-sensitive materials.
- Use of correct resolution for discriminating between adjacent anomalies.
- Dependence of resolution on scan speed, repetition frequency, etc.

7.5.4 Radiographic Testing

Radiographic testing (RT) uses electromagnetic radiations in forms of X-rays and gamma rays and is considered to be extremely capable of evaluating the presence, location, type, and size of internal discontinuities. For establishing the shape and location, the radiations are applied from two or three different directions on the components to be tested.

Radiographic test works on the simple principle. X-rays and gamma rays have the capability to penetrate or pass through the materials. These radiations are partially absorbed on passing through the materials. However, the extent of absorption depends on the density of material through which they are passed. The absorption is more in the case of high density of materials (like metal) than the low-density materials (like slag, air), and hence accordingly, the intensity of radiation reaching at the bottom of the surface is reduced. A film sensitive to these radiations is placed on the bottom of the component being tested. The film is affected as per the intensity of radiations reaching to the bottom. In case of sound metal system, radiations in the entire area of the component are absorbed uniformly, so the influence of the radiation on the film is almost the same, whereas in cases where the component in consideration contains discontinuities like cracks, porosity, inclusion, poor bonding at the interface, the film is affected at different levels in different areas.

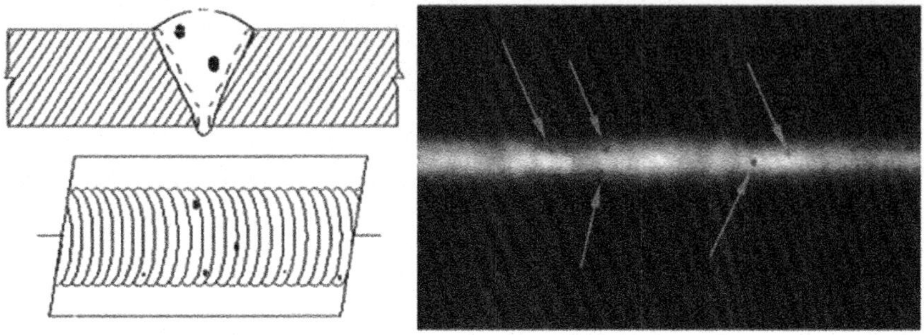

Fig. 7.14 Typical radiographic images showing defects [http://www.ndt.kalkars.com/radiographic-test-film-interpretation]

Locations and areas of component having discontinuities absorb less radiation; therefore, more radiation reaches at the bottom surface and affects the film to a greater extent, thus making those areas of film darker than other areas (Fig. 7.14).

Radiations used in the radiographic test are extremely harmful to the human being; therefore, proper protection and shielding arrangement must be made to avoid exposure of human being to these radiations using concrete and steel walls of suitable thicknesses.

Each test has unique advantages and limitations which dictate their applications as indicated in the Table (ASM handbook) below.

NDT test	Advantage	Limitation	Applications
DPT	• Simple • Cost effective • Portable	• Not for subsurface defects • Difficult to assess fine cracks • Surface cleaning is important	Surface discontinuities such as cracks, fine porosities
MPT	• Easy to apply • Quick • Simple	• Only for near surface defects • Only for ferromagnetic materials • Chances of arcing at contact point • Difficult to assess deep subsurface defects	Fine surface defects closed by impurities
UT	• Very sensitive method • Precisely locates the defects	• Difficult to interpret the results and accuracy depends on many factors • Needs expertise and skill to interpret findings	For both surface and subsurface defects like porosity, internal defects, etc.

(continued)

(continued)

RT	• Positive record of test is obtained • No limit on thickness of the material which can be evaluated	• Difficult to interpret the results and accuracy depends on many factors • Needs expertise and skill to interpret findings • Specially precaution is needed to handle radiations and protect operators	Internal defects can be located precisely

7.6 Destructive Testing of Modified Surfaces

Destructive tests such as hardness, tensile, toughness, fracture toughness, and other tests under simulated conditions are extensively useful for variety of purposes. In general, destructive tests are carried out to generate the data on mechanical performance of the specimen under investigation and to assess their suitability for the given service conditions. Additionally, destructive tests can also be used to identify or confirm the surface modification process used for the development of the engineered surface under study and to confirm if particular heat treatment has been performed properly. Hardness test is commonly carried out on small specimens for evaluating heat treatment, estimating ultimate tensile strength, and determining the extent of work hardening, or decarburization occurred on the component.

7.6.1 Hardness Testing

The hardness in surface engineering is more of qualitative importance than design purpose because it is used as an indicator of quality and ability of a modified surface to take up external load. Hardness indicates whether alloying, post-processing treatment such as heat treatment has been performed properly. However, hardness values are not used directly for any design purpose. It is always relative.

Resistance to indentation is refereed as hardness. Hardness of a material determines the depth of impression or indentation by a point object under a given load. High hardness results in low depth of indentation. The hardness of material indicates ability to withstand against abrasive wear conditions. Indentation is prerequisite for abrasion due to (a) presence of a hard particle between two mating surfaces, (b) two mating surfaces of high roughness, so soft surface is indented by asperities on hard surfaces, and (c) hard particle indents the soft surface. For removal of materials by abrasion, relative movement of mating surfaces or hard particles and surface of the component is required.

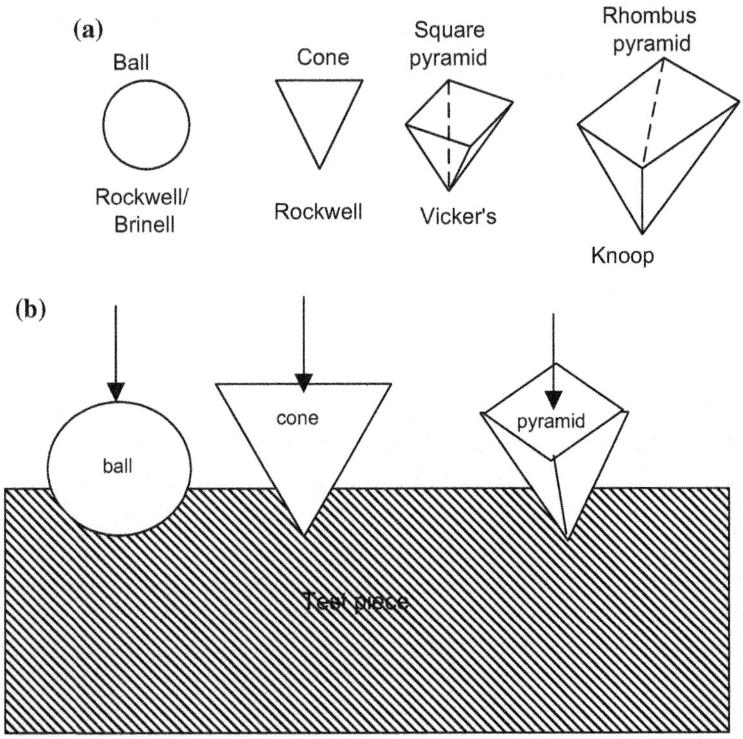

Fig. 7.15 Different types of **a** indenters and **b** pattern of indentation for various tests for hardness

In the most of methods used for the hardness testing of materials (a) first a standard load is applied using indenter of standard geometry and then (b) evaluating the effect of load on the surface of materials in form of impression/penetration. The penetration can be measured in terms of (i) diameter of impression, (ii) diagonal of impression, and (c) depth of impression (Fig. 7.15). High penetration suggests low hardness. In view of common principle of hardness testing method, these can be distinguished on the basis of the three parameters (1) indenter, (2) load, and (3) dimension of impression measured. Load for hardness of materials is generally taken as more than 30 N and that for microhardness ranges from 10 mN to 10 N.

Parameter	Brinell	Rockwell	Knoop	Vickers
Load	500–2000 kg	Minor: 10 kg Major: 40–200 kg as dictated by scale to be used (A–C)	10–3000 g	
Indenter	Steel ball	Steel ball or diamond cone	Diamond cone	Diamond pyramid
Measurement	Diameter	Depth	Diagonal	Diagonal

The size of indentation during hardness test is influenced by surface condition such as surface irregularities and impurities. Presence of hard oxide layer on the surface of samples can reduce size of indentation which in turn may incorrectly show high hardness. Cleaning of surface and removal of surface irregularities from the test sample is considered as good practice for hardness test. However, effectiveness of surface irregularities and presence of impurities on accuracy of hardness test results vary with test method being used. Hardness test methods (Brinell hardness testing) where standard load for indentation is applied on the specimen directly, accuracy of test results may be affected more by surface impurities and irregularities than those methods (Rockwell) where first minor load is applied to ensure good metallic contact between the indenter and specimen. Minor load applied through indenter breaks the surface impurities oxides and irregularities to enhance the metallic intimacy between indenter and specimen. Minor load is not expected cause any penetration. Penetration is caused by standard major load only.

Brinnel hardness test uses a steel ball (of diameter, D) as indenter and average diameter of indentation (d) to evaluate the hardness. Hardness values shown by Brinell test shows pressure developed due to standard load (P) over the projected area of indentation and obtained from the relation between standard applied load and projected area of indentation. High standard load is selected for test hard material.

$$\mathrm{BHN} = \frac{2P}{\pi D[D - (D^2 - d^2)]^{1/2}}$$

The standard major load in case of Rockwell hardness test can vary from 40 to 140 kg as per scale A, B, C, D to be used for hardness test, while the minor load (10 kg) remains same. The selection of scale for hardness test depends on the specimen material. B, and C scales are commonly used for hardness test of most commercial metals. Indenter of ball and cone shape made of steel and diamond, receptively, used B and C scale respectively (Fig. 7.16). Standard load for B and C scale is 90 and 140 kg, receptively, for Rockwell hardness test.

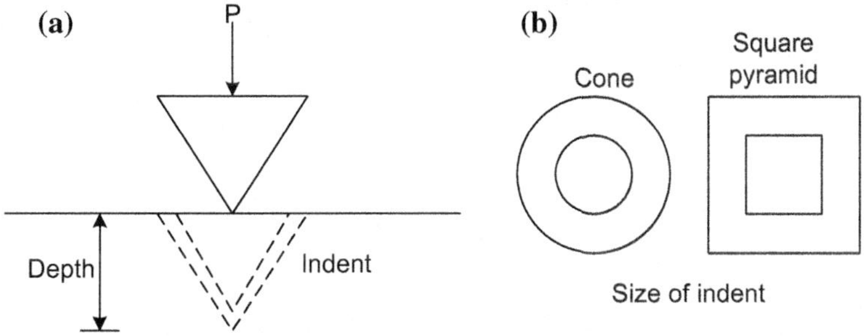

Fig. 7.16 Schematic of conical and square pyramid indenters **a** front view and **b** top view of indentation

A pyramid of square base made of diamond is used as indenter for Vickers hardness testing. Standard load for this testing varies from 1 to 120 kg. Indentation geometry formed on specimen surface becomes square. Average length (*L*) of diagonals of square is used to calculate the Vickers hardness. High value of diagonal of square indentation shows low value of Vickers hardness number (VHN). Vickers hardness is calculated from the ratio of standard load (*P*) and apparent area of indentation using equation given below. Similarly, Knoop hardness number (KHN) is also calculated (Fig. 7.17).

$$\mathrm{VHN} = 1854P/d^2\{\text{where } d \text{ is the average length of diagonal } d = (l_1 + l_2)/2\}$$
$$\mathrm{KHN} = 14229P/d$$

7.6.2 *Microhardness*

Microhardness is used for measuring the hardness of individual microconstituents by using very fine indenter and low normal load so that indentation is limited to modified region (coating) and does not have influence of the substrate hardness (Fig. 7.18). For example, hardness testing of different micro-constituents in flame sprayed coating, Ni–WC composite materials, having hard constituent WC of 2500 HV is converted into W_2C of 2100 HV hardness. The extent of thermal damage can be determined by measuring the hardness of different constituent and grains individually as the function of distance from the surfaces or coating-substrate interface. As coating wears out gradually during service, the wear resistance of the coating also varies owing to variation in the hardness of the subsurfaces region (Fig. 7.19).

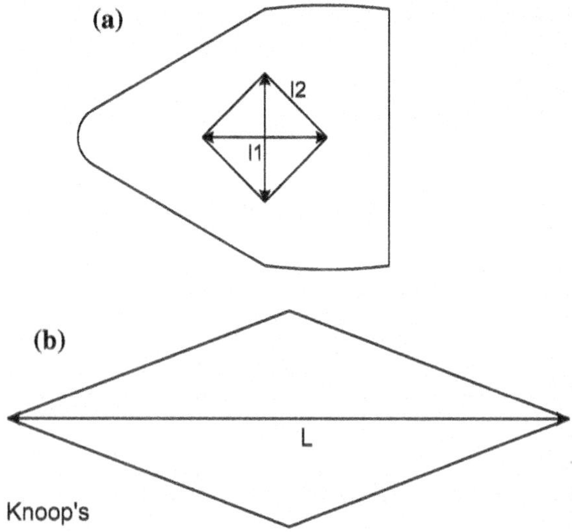

Fig. 7.17 Schematic of indentation of hardness tests **a** Vickers hardness indentation and **b** Knoop hardness indentation

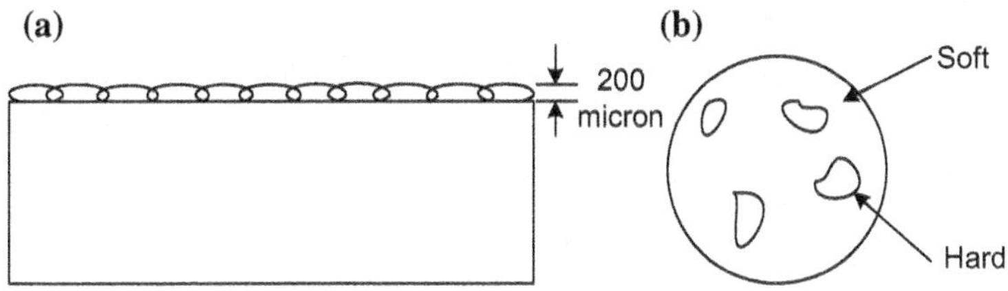

Fig. 7.18 Schematic of **a** coating–substrate system and **b** microconstituents in matrix of the coating

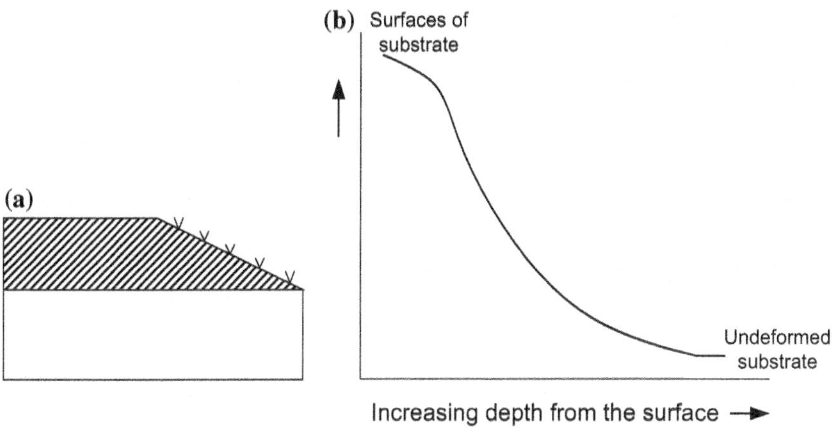

Fig. 7.19 Schematic of microhardness testing approach for coating and near surface layers using **a** slant surface preparation and **b** hardness variation of subsurface region as function of increasing depth from the surface

7.6.3 Tensile Test

The tensile test is one the most common test to evaluate (a) the quality of heats and (b) general mechanical load carrying capacity. Tensile test gives value of various parameters (elastic limit, modulus of elasticity, yield strength, proof stress, ultimate strength, elongation at fracture, yield to ultimate strength ratio, etc., indicating the response of specimen under tension (Fig. 7.20). As per the service requirement of an application, the tensile test can be performed in different environments such as high temperature, low temperature, corrosion, erosion. This test is usually conducted at constant strain rate. It is mandatory to include following details of test while reporting the results:

- The direction wherefrom specimen was taken (transverse, longitudinal, all weld, etc.) from bulk material.
- Strain rate at which test was conducted (0.0001–10,000 mm/min).

Fig. 7.20 Schematic of typical stress–strain diagram for stainless steel sample **a** engineering stress–strain diagram and **b** method of determining the proof stress using stress–strain diagram

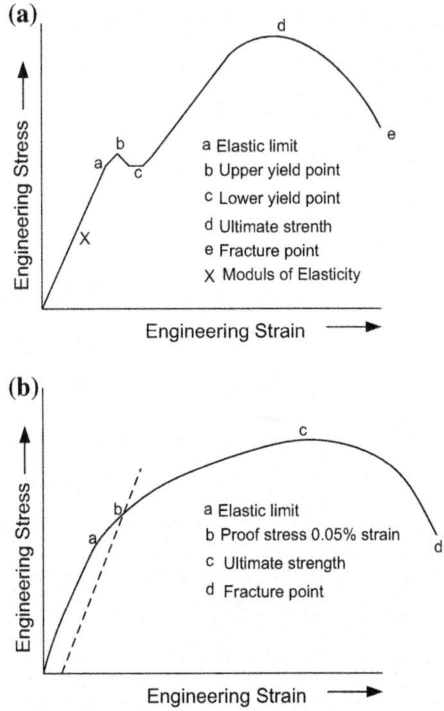

- Temperature: room temperature or any specific high or low temperature.
- Specific environment if any: ambient condition, corrosion, or erosion
- Morphology of the fracture surface
- Mode of fracture
- Stress state
- Fracture location.

7.6.4 Toughness Testing

In actual practice, engineering components during service are invariably subjected to various kinds of loads, namely static and dynamic loads, which are classified on the basis of the rate of change in magnitude of load and direction. Dynamic loads are characterized by high rate of change in load magnitude and direction. Reverse happens in the case of static loads. For the hardness test and tensile tests, load is increased very slowly and this corresponds to the behavior of material under more or less static loading condition. Moreover, very wide range rate of loading (0.0001–1000 mm/min) can be used in the tensile test. The rate of loading governs the strain rate, and so rate of hardening, which in turn can affect mechanical behavior of

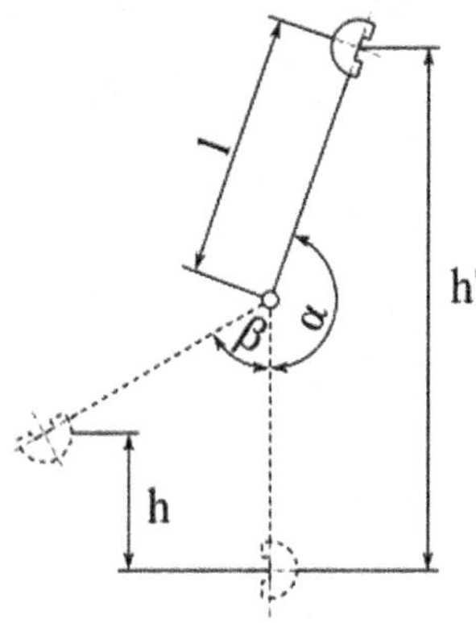

Fig. 7.21 Schematic showing principle diagram of toughness test

material. For example, material at low rate of loading showing the ductile behavior can exhibit brittle behavior under high rate of loading conditions.

This test simulates service conditions which often encountered in the transportation, agricultural, and construction equipment. A material which possesses high impact resistance is said to be a tough material. Toughness is the ability of a material to resist both fracture and deformation. Toughness is the combination of strength and ductility. To be tough, a material must be both fairly strong and ductile to resist cracking and deformation under impact loading. Notches are placed in the impact of test specimens to increase the stress concentration, so as to increase the tendency toward fracture because mostly mechanical components have stress raisers due to various geometrical requirements (Fig. 7.21). To withstand an impact force, a notched material must be tough.

To study the behavior of material under dynamic load conditions (at high rate of loading), toughness test is frequently conducted. There are two methods used for toughness testing, namely Izod and Charpy test. In both methods, load is applied at high rate on to the standard test specimen and the amount of energy absorbed (N.m or J) for breaking the sample due to impact is measured (Fig. 7.22). However, there are some differences in these two methods in terms of sample size and shape, method of holding of the sample, and maximum energy content of pendulum that hits the sample during the test.

S. No.	Toughness test	Sample	Holding
1	Izod	Held vertically on anvil as cantilever	Cantilever and notch faces the pendulum
2	Charpy	Held horizontally on anvil as simply supported beam	Simply supported and notch is on the opposite side (not facing to pendulum)

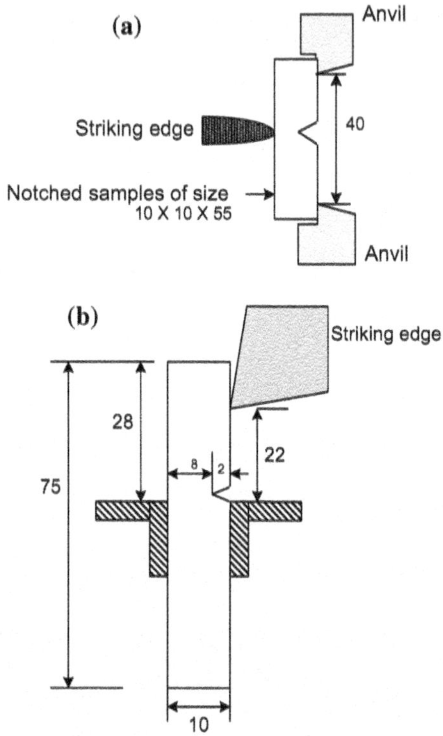

Fig. 7.22 Standard specimens for **a** Charpy and **b** Izod impact test

Standard sample for both testing methods having a notch and is mounted on the machine in specific ways, that is, notch faces to pendulum in case of Izod test, while pendulum hits the sample from back of the notch in the Charpy test (Fig. 7.22).

Since most of the engineering components are invariably designed with notch and stress raisers, therefore, it becomes important to know about the behavior of material under impact loading in notched condition. Hence, the toughness test is usually conducted using sample with notch. Moreover, un-notched samples can also be used for the toughness test and the results are expressed accordingly.

The results of impact tests are expressed in terms of either the amount of energy absorbed or the amount of energy absorbed per unit cross-sectional area. It may be noted that the values of toughness are not directly used for design purpose, but these only indicate the ability of the material to withstand against shock or impact load, that is, load applied at a very high rate. These tests are useful for comparing the resistance to impact loading of different materials or same material with different processing conditions such as heat treatment, procedure, and mechanical working. The resistance to the impact loading of material depends on the surrounding temperature (Fig. 7.23). Therefore, temperature at which toughness test is conducted must be reported with results.

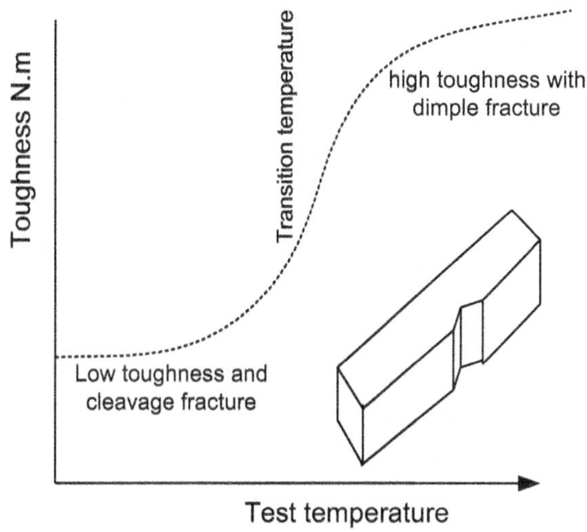

Fig. 7.23 Schematic diagram showing the influence of test temperature on toughness

7.6.5 Fatigue Behavior

The fatigue performance of the metallic components in general is determined in two ways: (a) endurance limit indicating either the maximum stress/stress amplitude or stress range for infinite life (typically more than 20 million of load cycles) and (b) number of load cycle a component can withstand for a given number of loading conditions as desired. Two types of samples are generally prepared for fatigue studies as per ASTM 466 (Fig. 7.24 a, b). Reduced radius sample generally ensures fracture from specific loaction (Fig. 7.25). The fatigue performance is appreciably influenced by the different variable associated with fatigue test, namely stress ratio, type of stress (tension–tension, reverse bending, tension compression, zero tension), maximum stress, stress range, loading frequency, and surrounding environmental conditions such as temperature, corrosion, vacuum, tribological conditions. Each and every parameter to be used for the fatigue test must be carefully selected and recorded with the results while reporting.

- Test conducted according to ASTM E466 standard.
- Type of loading: axial pulsating/reverse bending/tension–compression.
- Maximum stress
- Stress ratio (ratio of minimum stress to maximum stress)
- Temperature: ambient/vacuum/corrosion
- Frequency of pulsating load: load cycles per min.

For plotting the stress vs number of cycle (S-N) curve, the fatigue test is first conducted with maximum applied tensile load corresponding to 0.9 time of yield strength of sample under study for determining the number of load cycle required for fracture and then the same test is repeated at 0.85, 0.8, 0.75, 0.7 … times of yield strength of weld joint until the endurance limits or desired fatigue life is

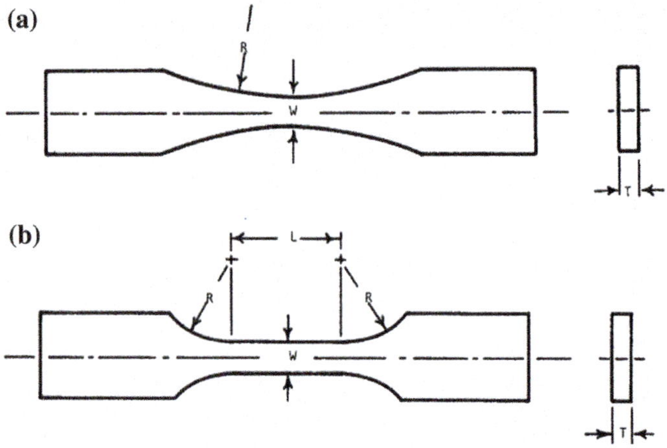

Fig. 7.24 Standard specimen for fatigue testing **a** reduced radius sample and **b** sample with a gauge length

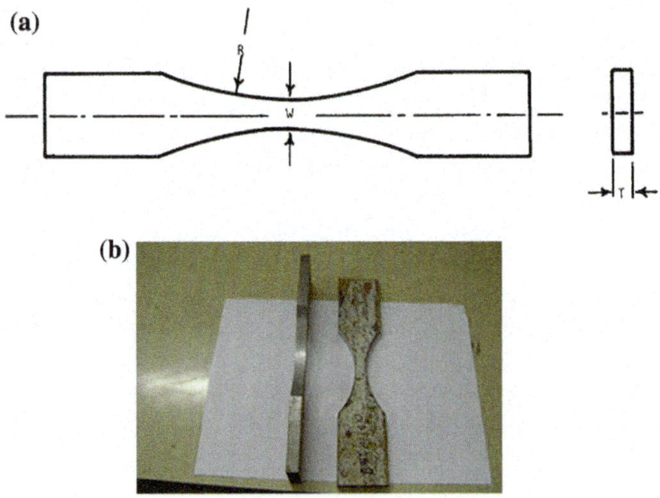

Fig. 7.25 Fatigue test sample **a** schematic diagram of standard fatigue test sample with continuous radius between ends and **b** photograph of typical specimen

achieved (Fig. 7.26). Typical dimensions of a standard specimen as per ASTM 466 are as under.

- Continuous radius (R): 100 mm
- Width (W): 10.3 mm
- Thickness (T): 11 mm (as received)
- Gripping length: 50 mm.

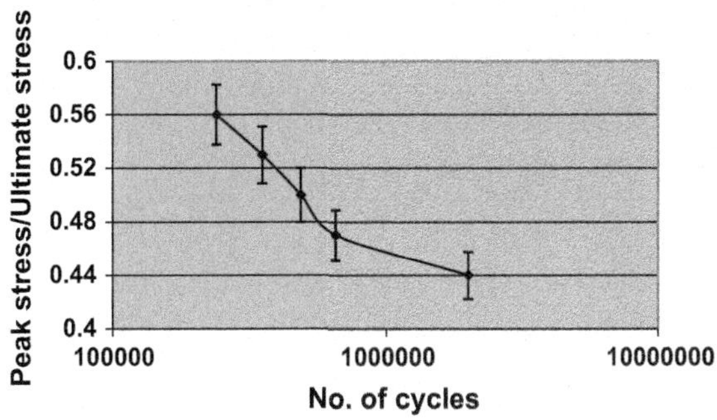

Fig. 7.26 Typical data on fatigue test showing peak stress/ultimate stress versus number of cycle relationship for structure steel

7.6.6 Fracture Toughness

The resistance to fracture conversely resistance to crack growth is known as fracture toughness and is measured using various parameters such as stress intensity factor around the crack tip (K), opening of crack mouth also called crack tip opening displacement (CTOD), and energy requirement for growth of crack (J or G). The mechanical properties, namely yield strength, ductility, and thickness of the specimen under study primarily dictate the suitable parameter to be used for determining the fracture toughness. The fracture toughness parameter, namely stress intensity factor (K) is commonly used for component of heavy sections of high strength and low-ductility material developing plain strain conditions, while crack tip opening tip displacement (CTOD) and energy-based methods (G and J integral) are used for comparatively thinner sections that are made of low strength and high ductility material resulting in plain stress condition. The measurement of fracture toughness using any of the above parameters is conducted using two types of samples: (a) compact tension specimen (CT) and (b) three point bending specimen (TPB). The schematics of two types of specimen are shown in Fig. 7.27a, b. In general, in these tests, the applied external load increase until strain or crack opening displacement or energy versus load relationship becomes nonlinear. This critical value of load (P) is used for calculations of fracture toughness using relevant formulas.

Although different standards have historically been published for determining K, CTOD, and J-integral, the tests are very similar and generally all three values can be established from one single test.

In general, stress intensity factor (K) decreases with increase in specimen thickness. This trend continues up to a limit of thickness, and thereafter, K becomes independent of the plate thickness. The corresponding value of K (which is independent of thickness) is called critical stress intensity factor (Kc) and occurs in *plane strain condition*. K_{IC} is used for the estimation of the critical stress to be applied to a specimen with a given crack length to cause unstable crack growth and so fracture.

$$\sigma_C \leq K_{IC}/(Y(\pi a)^{1/2})$$

Fig. 7.27 Schematic of fracture toughness specimens using **a** compact tension and **b** three point bending test sample. $W = 2B$, $a = B$, $W - a = B$ and radius of hole $r = 0.25B$, where B is plate thickness

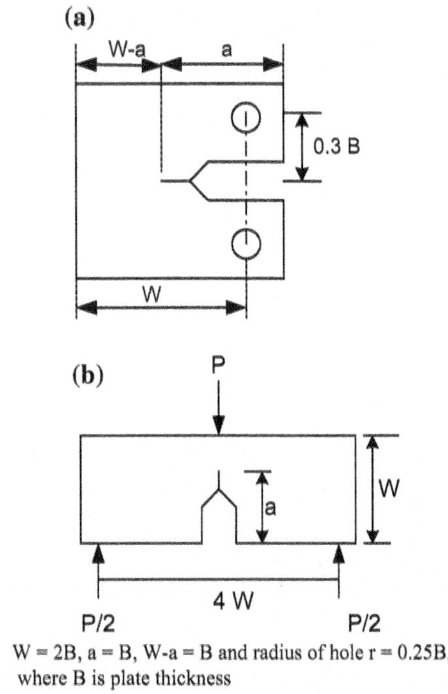

where K_{IC} is the stress intensity factor measured in MPa * $m^{1/2}$, σ_C is the critical stress applied to the specimen, a is the crack length for the edge crack or half crack length for internal crack, and Y is a geometry factor.

7.7 Adhesive Wear

To evaluate the adhesive wear resistance of metal and engineered surfaces depending upon the application and loading conditions, various wear test configurations such as pin on disk (POD), pin on ring, pin on block, ball on plate are used for performing adhesive wear. The adhesive wear test based on all these configurations are performed under unidirectional sliding conditions. Therefore, the localization of frictional heat generated at the contact interfaces during sliding of test specimen and counter surface is governed by normal load affecting the frictional force, relative speed, dimensions of wear test specimen and counter surfaces and their thermal conductivities. Since the localization of heat at the interfaces influences the thermal softening and oxidation tendency of wear specimen, therefore, all factors affecting it must be carefully considered as test parameters and these must be specified with results. Important parameters of adhesive wear tests include normal load, sliding speed, counter surface (hardness, surface roughness, thermal conductivity, and metallurgical properties), interfacial contact conditions like lubrication, oxidation, besides contact geometry like point contact, line contact, and area contact. In the following section, the typical methodologies for adhesive wear test for pin-on-disk approach and pin-on-block approach have been described.

7.7.1 *Procedure of Pin-on-Disk Unidirectional Sliding Adhesive Wear Test*

A pin-on-disk friction and wear testing machine with suitable data acquisition system if any (Fig. 7.28) can be used to evaluate the friction and wear behavior of the modified surfaces in consideration against suitable counter surface (hardened ground steel made of En-31 with composition Fe-2.3% Cr-0.9% C and hardness 60 HRC and surface roughness (R_a) of the order of 0.3 μm) or any other suitable counter surface as per need of application. Wear test pins of suitable diameter (5–10 mm) and length (20–30 mm) can be prepared by machining process, and one end of the pin is abraded with polishing paper (R_a–0.2 μm). The counter surface disk is rotated with suitable electric motor (say DC motor) having required rotational speed (10-2000 rev min^{-1}) with required wear track (diameter 20–180 mm), so as to obtain desirable sliding speed (0–10 m/s).

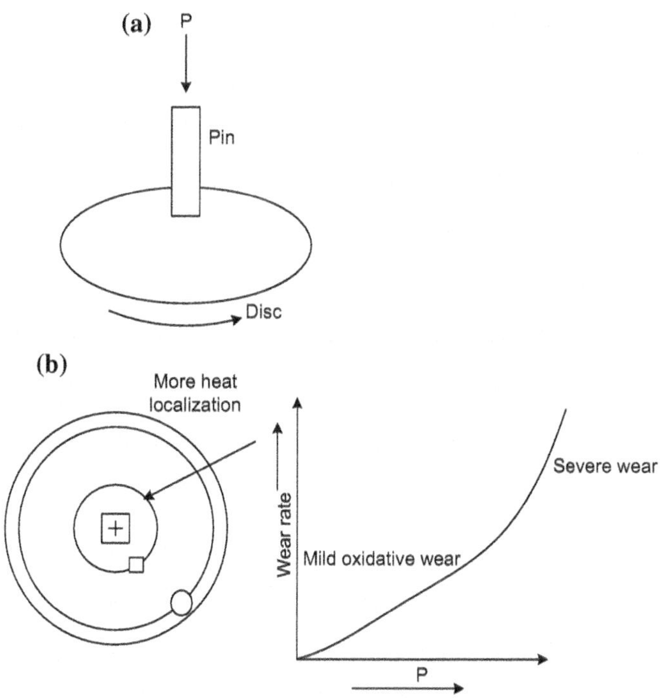

Fig. 7.28 Schematic of **a** pin-on-disk arrangement and **b** wear versus load relationship

Normal load is applied on the pin sample by suitably using dead weights through a pulley–string arrangement. Counter surface is polished with polishing paper, cleaned with acetone, and dried before each sliding test to maintain identical sliding conditions. Temperature and wear data can be acquired during sliding. Sliding distance traversed for each test can be fixed such that the steady state of wear is achieved. Wear test under each condition should be repeated thrice to have more representable wear behavior.

Temperature of the wear pin near the subsurface region during sliding can be measured using suitable method such as thermocouples. This method is very easy and convenient to measure the near wear surface temperature. It considers the heat transfer through conduction only and neglects the losses of heat by radiation. Thermocouples can be placed into holes each of 1 mm diameters at suitable intervals (such as 1.5, 3.0, and 4.5 mm) away from sliding surface. Holes are drilled up to the axis of cylindrical pin. To find out the temperature at the sliding surface, the extrapolation method is used in this approach. Wear pin temperature at the sliding surface can be useful in understanding and explaining the friction and wear behavior because mechanical properties of material and surface oxidation are affected by surface temperature.

7.7.2 Reciprocating Pin-on-Flat Wear Test

Wear behavior of modified surfaces can be studied under reciprocating condition using pin-on-flat configuration for applications like piston, where mating/interacting components are subjected to reciprocating sliding conditions (Fig. 7.29). Under reciprocating sliding conditions, the wear rate found is generally higher than that under unidirectional sliding conditions because of the presence of wear debris on the wear tracks, which may act as a third body to cause abrasive wear. Therefore, it may be feasible to test the specimen under reciprocating sliding condition as per the need of application. Most of the parameters of reciprocating pin-on-flat sliding wear test are same as that of unidirectional pin-on-disk sliding wear test, except variation in speed during sliding (zero at both the ends and maximum at the center), length of wear track, and frequency of reciprocation. The procedure of a typical reciprocating pin-on-flat sliding wear test is presented in the following section.

The wear pin of suitable size (e.g., $5 \times 5 \times 20$ mm^3) is held against flat plate (hardened steel or cast iron of size $100 \times 50 \times 10$ mm^3), used as counter surface during wear test. The counter surface is reciprocated using suitable mechanism and desired reciprocating average sliding speed (e.g., 0.5 m/s) is attained as per the requirement of test for specific application. Sliding distance (e.g., 1000 m) as per sliding condition (load, speed, counter surface, etc.) is selected in such a way that the steady state of wear is achieved. Sliding wear tests may be conducted at different normal loads (5-20 N). A suitable microbalance of high accuracy (>0.0001 g) is used for weighing the specimens before and after the sliding. Weight loss can be used as a measure of wear. Wear rate is calculated using weight loss per unit sliding distance. The wear pins are cleaned with acetone prior to and after the wear test.

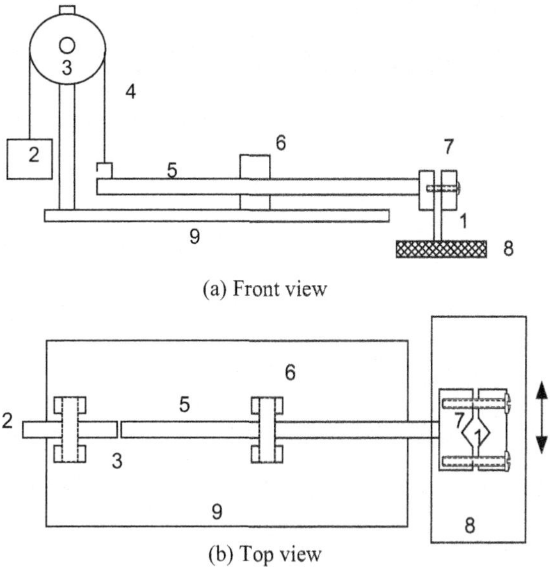

Fig. 7.29 Schematic diagram of pin-on-plate wear test machine (1 Wear pin, 2 weight, 3 pulley, 4 string, 5 lever arm, 6 fulcrum of lever, 7 wear pin holder, 8 reciprocating counter surface affixed with abrasive medium, and nine base plate)

7.7.3 Abrasive Wear Test

Abrasive wear test rigs

Various abrasive wear testers have been used around the world and some of them are standardized with well-defined experimental procedures. The American Society for Testing and Materials (ASTM) has tried to standardize the most popular wear testers. Few standardized abrasive wear testers are (i) pin on disk, (ii) pin on drum, (iii) pin on belt, and (iv) Pin on flat (Figs. 7.30 and 7.31).

In pin-on-disk wear testing, the disk carrying the abrasive medium (abrasive paper) is rotated at a fixed rpm (revolution per minute), while the motion of the pin can be circular or spiral. In case of pin on drum, the abrasive medium in the form of abrasive paper is wrapped around the cylindrical drum, while the pin is moved at a constant speed in a straight line, thus, the resultant motion of pin on drum is helical. Similar is the case for pin-on-belt tester. The abrasive belt is moved

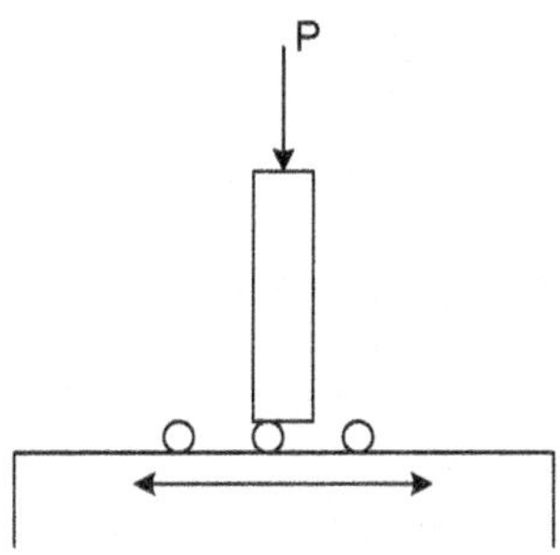

Fig. 7.30 Schematic diagram of pin-on-flat wear test machine

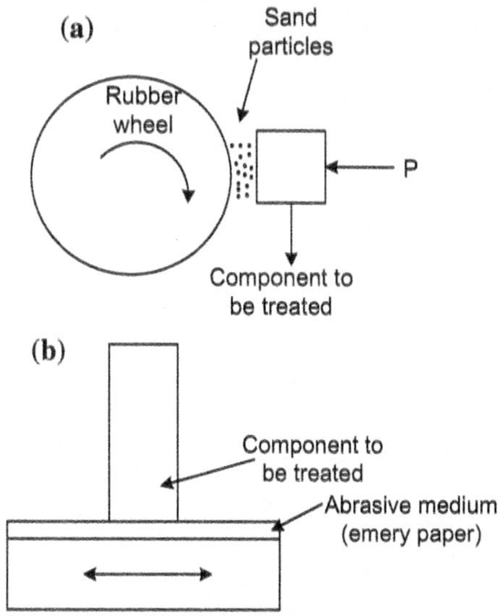

Fig. 7.31 Schematic diagram of **a** rubber wheel abrasive wear test and **b** pin on flat

at fixed rpm, while the pin is moved at a fixed speed linearly in each rotation of the belt. Thus, the relative motion between the pin and belt is helical. In the case of pin on flat, the abrasive paper is mounted on the flat. The pin is held against the flat, and the flat carrying the abrasive paper is reciprocated. At the same time, after each pass of the pin, the flat is moved crosswise so that fresh abrasives come in contact with pin after each pass. The parameters of abrasive wear test include normal load, sliding speed, type and size of abrasives, and temperature.

The pin-on-disk type abrasive wear tester has been described in the following section. During test, the wear pin is held against the rotating abrasive medium of different sizes mounted on a steel disk and spiral motion was obtained by moving the wear pin radially outwards.

Abrasive wear can be measured in terms of weight loss, volume loss, and depth of the worn surfaces. To overcome the issues of actual contact area and roughness of the worn surface, the weight loss approach is commonly used on a measure of abrasive wear. The measurements of volume loss in field tests often fail to give reproducible values.

Abrasive wear test (two-body) can be carried out using the pin-on-disk wear and friction monitor. Rectangular shape specimen having suitable dimensions ($6 \times 6 \times 40$ mm^3) is cut from the modified surfaces. The samples are ground to remove the sharp edges before the sliding tests. Polishing paper embedded with the desired size of SiC particles (320 grade) can be used as abrasive medium. The specimen is fixed with a locking arrangement against the abrasive medium. Load is applied on the specimen by cantilever method. The sample is made to slide under suitable normal loads (2–10 N) at a constant sliding speed (0.2-1.0 m/s) against the abrasive medium fixed on the counter surface. A microbalance (accuracy 0.1 mg) can be used for weighing the specimens before and after the sliding. Weight loss is used as a measure of wear. Wear rate is calculated using weight loss per unit sliding distance. The specimen is cleaned with acetone prior to and after the wear test. Each test is conducted with fresh abrasive medium. Sliding tests are carried out for (200 m) sliding distance. Wear rate is calculated for weight loss measurement after specific interval (e.g. 50 m) sliding distance.

7.7.4 *Solid Particle Erosive Wear Test*

There are a number of designs of erosion test rigs used to study the erosive wear behavior. The four major type of test rigs are: (i) the sand or gas blast rig, (ii) reciprocating type erosion tester, (iii) centrifugal accelerator, and (iv) whirling arm rig. In the gas blast rig, particles are introduced into a fast moving gas stream and accelerated down the acceleration tube before impacting with the test piece. In the slurry type, gas stream is replaced by water stream. In these test rigs, the motion of gas and water stream is linear, whereas, in the case of centrifugal accelerator and whirling arm rig, the motion is circular. In the centrifugal accelerator, the test samples are held stationary at the rim of the rotor and the sand particles are fed into the center being accelerated radially impact on to the target surface. The test duration is often long. In whirling arm rig, the samples are mounted onto the end of the arms, which are attached to a rotor. The rotor is rotated at a high speed, and erodent is fed on to the target surface.

For solid particle erosion testing at high temperature, the test rig is composed of an air compressor, particle feeder, an air–particle mixing and accelerating chamber, furnace unit, pneumatic and electrical controls. Dry compressed air mixed with the particles is fed at a constant rate from a conveyor belt-type feeder in the mixing chamber and then accelerated by passing the mixture through a converging nozzle (Fig. 7.32). These accelerated particles affect the specimen, which could be held at various angles using an adjustable sample holder. The feed rate of the impacting particles (erodent) is controlled by monitoring the distance between the particles feeding hopper and the belt drive that carry the particles to the mixing chamber. The motor speed determines the belt speed. Higher the motor speed, higher is the

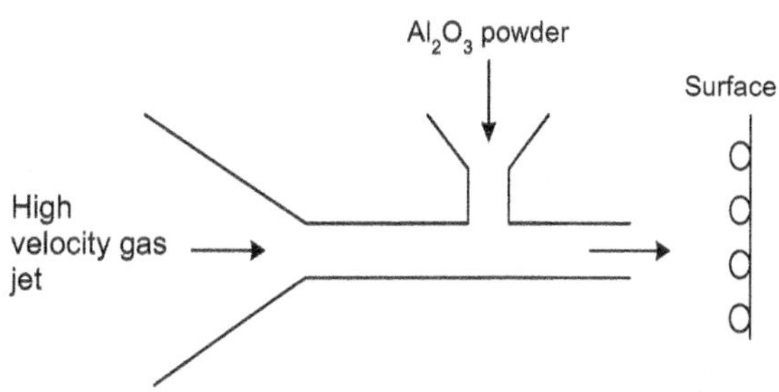

Fig. 7.32 Schematic showing principle solid particle erosion test

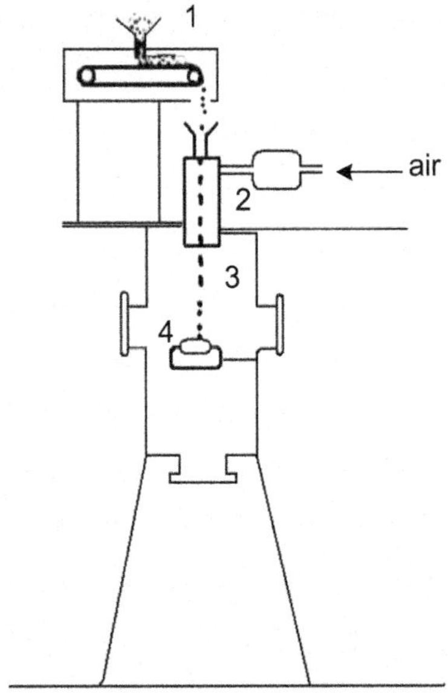

Fig. 7.33 Schematic of solid particle erosion testing machine showing 1 erosive powder feeder, 2 powder-compressed gas mixture 3 erosion chamber, and 4 test specimen

erodent feed rate. The velocity of the impacting particle can be determined by a rotating double disk method.

Typical procedure of erosive wear test

The samples are cut into suitable size (16 × 15 × 5 mm^3) for erosive wear testing. These are polished to 1–2 μm surface roughness (Ra) in order to obtain identical surface conditions on all erosive wear samples. The samples are cleaned with acetone, dried, and weighed to an accuracy of 0.0001 g before erosive wear test.

The impact velocities of the particles can be varied by changing the pressure of the compressed air (Fig. 7.33). As per ASTM G75-96 (2000), the standard erosion value is obtained from volume loss of specimen material divided by the total mass of abrasive particles that affected the specimen ($mm^3\ g^{-1}$). Normalized erosion rate is calculated by dividing the erosion value of the specimen by erosion value of the reference material. The erosion rate is defined as the ratio of the change in the mass of test sample and the mass of the impacting particles. The erosion resistance is also expressed as the time required to erode 25.4 μm (1 mil) of sample material (coated and uncoated), which is determined by measuring the average depth of the pit produced during the test. Data for coated specimens are depicted on the basis of the time required for the erosive material to penetrate the coating on a per mil basis. Higher the number, the better is the erosion resistance of the coating. The penetration time is determined by repeating the test at least three times for the coated specimens to check repeatability of the test results. The variation in penetration time is less than ±5% from the mean value, indicates good repeatability of the test results.

7.7.5 *Slurry Erosive Wear Test*

Solid hard particles (like sand) are mixed with water to form slurry in varying concentration. Sand particles are allowed to impact the test specimen at high velocity. Relative velocity among the solid particles and test specimen surface is achieved in different ways: (a) by rotating the samples held in a suitable sample holder in a chamber having slurry of desired concentration of sand particles in water and (b) by directing the high velocity jet of water carrying sand particles on the surface of specimen to be tested. The repeated impacts of sand particle with jet on the surface of specimen result in the removal of material by erosion. Loss in weight of test specimen by conducting the test after a specific time (e.g., 30 min) is used as a measure of wear. The concentration of sand particles in slurry, speed jet, or rotational speed of test specimen, type, and size of sand particles and test duration are few parameters of slurry erosion test.

Typical procedure of slurry erosion pot test
The slurry erosion pot test equipment essentially comprises of a steel container for holding the slurry of silica sand (e.g., average particle size of 452 μm) and water. Specimens are rotated in the slurry (Fig. 7.34). The specimen of rectangular shape ($12 \times 12 \times 35$ mm^3) is held in specimen holders, fitted to a mild steel bar at some radial distance, (100 mm). The steel bar is attached to an electric motor assembly which is capable of rotating the specimens at a desired speed. The samples are thoroughly ultrasonically cleaned and weighed prior to and after the erosion tests. Weight loss is used as a measure of erosion. Electronic balance of 0.1 mg accuracy or better is used to measure the weight of specimen. Erosion test is conducted by traversing the coated specimen for enough time, for example, one hour in slurry of varying concentration (20 and 40% sand) at different rotational speeds (600, 800, and 1000 rpm).

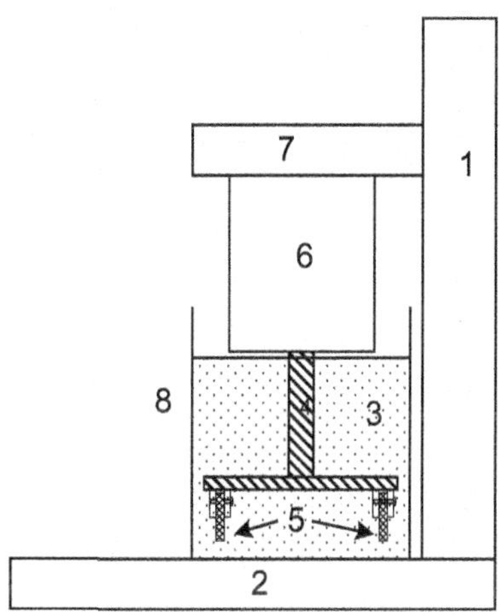

Fig. 7.34 Schematic diagram of the slurry pot tester, 1 column, 2 machine base, 3 slurry, 4 spindle, 5 sample, 6 motor, 7 motor base, and 8 slurry tank

7.7.6 *Cavitation Testing Rigs*

The resistance to cavitation of the modified surfaces is evaluated by developing flow conditions which are suitable for creating cavitation and putting in samples to be tested close to the cavitating zone. However, the conditions suitable for producing cavitation bubbles are developed using various methods. The commonly used method is creating a low-pressure region mostly by developing disturbance in the fluid flow system. The disturbance is created using the following three common methods: (a) ultrasonic vibration near the test specimen submerged in liquid, (b) flow of liquid through a venturimeter, and (c) rotation of wheel (having various geometrical features in the forms of holes) in a closed chamber filled with liquid. Bursting of bubble develops shock waves of high pressure, which in turn causes the loss of materials from the specimen surface. Weight loss, number of pits, average depth of pits, and surface roughness of the tested surface are some of the features that are used as a measure of cavitation wear.

Vibratory horn cavitation erosion test apparatus
The entire test setup consists of two subsystems, which are sonifier and sound enclosure. Sonifier generates the required ultrasonic vibratory movement for cavitation. It is composed of four elements, namely power supply, controls, converter, and horn. The power supply of sonifier converts AC line voltage to 20 kHz electrical energy. This high-frequency electrical energy is fed to a converter. The converter is a magneto-strictive device, which expands and contracts with the supply of high frequency AC voltage. This expansion and contraction appears in the form of longitudinal vibrations, which are transmitted from the convertor to the horn tip immersed in liquid (Fig. 7.35). The vibratory motion of constant amplitude at horn tip generates cavitation bubbles. The ultrasonic vibrations can be applied directly to the surface of specimen through suitable tips attached to the horn. The

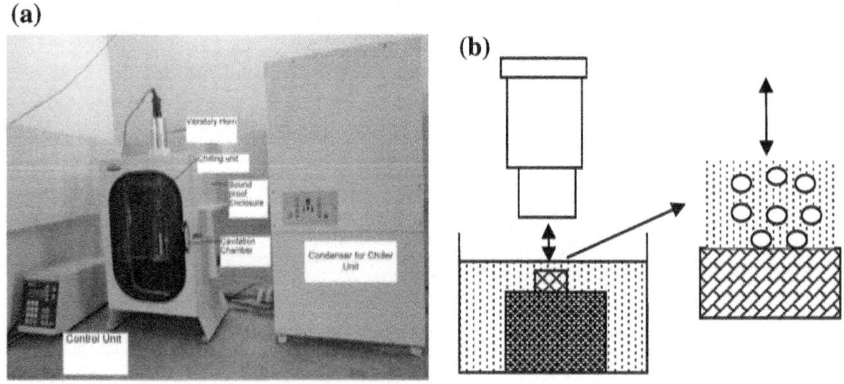

Fig. 7.35 Vibratory horn cavitation test rig **a** photograph of the set and **b** principle of test

sound enclosure minimizes the mechanical noise (20–25 dB) produced by ultrasonic vibrations.

Typical procedure of cavitation erosion testing

Cavitation erosion test can be performed using suitable vibratory horn, and general methodology described in ASTM G-32 is as under (Fig. 7.35). The diameter of the horn tip is about 12.7 mm, and therefore, the diameter of the test specimen surface is chosen (of 12.7 mm size) to match the surfaces. The shank of each specimen can be 10 mm in length. The thickness of each type button is kept 3.56 mm with suitable arrangement to facilitate tightening (and removing) of the specimen to the steel base facing against the horn tip. Specimens are cleaned with tap water, rinsed in acetone, and dried with hot air blower before and after each test. Cavitation testing is performed by tightening the specimen to stainless steel holder, and this holder with specimen is placed in a glass beaker, which is filled with distilled water. The typical experimental parameters are given in the table below. Mass loss of the wear specimens after each test can be measured by an electronic balance (least count of 0.1 mg).

Typical parameters for cavitation erosion testing

Type of investigation	Value
Peak to peak amplitude	50 μm
Vibrating frequency	20 kHz
Run time (h)	3 (measuring weight loss after each 30 min interval)
Test liquid	Distilled water
Temperature of test liquid	25 ± 2 °C

7.7.7 Corrosion Testing

The corrosion behavior of modified surfaces can be investigated by measuring the variation in corrosion potential and current density of potentiodynamic polarization curves for carrying out the Tafel analysis. The open circuit potential (OCP) or corrosion potential (E_{corr}) is a unique mixed potential at which the rates of anodic and cathodic reaction are exactly same and are equal to the corrosion rate.

Potentiodynamic polarization curves for Tafel analysis (in 3.5% NaCl solution) are shown in Figs. 7.36 and 7.37. The cathodic current density shows the corrosion rate. The values of breakdown potentials are generally found to be in good agreement with the OCP for all the samples tested as per their corrosion behavior. Low corrosion potential (more negative) value indicates poor corrosion resistance. Accordingly, it can be observed from Fig. 7.37 that base metal having better corrosion resistance than weld nugget and heat affected of friction stir weld joint of AA 2024 alloy.

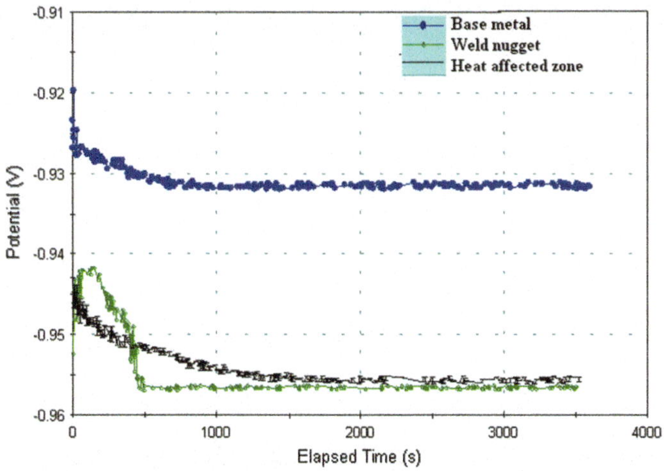

Fig. 7.36 Potentiodynamic polarization curves for Tafel analysis

Procedure for immersion corrosion tests

Immersion corrosion tests is performed (in accordance to ASTM G3-89) on the specimens having very good surface finish (polished to 1200 grit polishing paper) and cleaned (with ultrasonic cleaner by using distilled water as a medium) and dried in air. Three electrodes are utilized in a flat corrosion cell (a) test sample of 0.5 cm^2 exposed area, (b) Ag/AgCl-saturated KCl reference electrode, and (c) platinum-wired counter electrode. Scan rates (0.166 mV/s) are applied to the cell (with the use of PARSTAT 2273®, and operated by the software PowerSuite®). The open circuit potential (OCP) measurement and potentiodynamic polarization experiments are carried out in 3.5% NaCl solution for investigating corrosion behavior of selected area. FE-SEM is used in backscattered electron mode for studying the corrosion affected surface after carrying immersion corrosion test in 3.5% NaCl solution. The FE-SEM is used for chemical analysis of corrosion products formed on surfaces by conducting EDAX analysis of the corroded surface and XRD is used for identification of compound/phases of corrosion product. The surface roughness after cyclic polarization experiments in 3.5% NaCl solution can be measured by scanning with atomic force microscopy probe (NT-MD, NTEGRA).

7.8 X-Ray Diffraction (XRD) Analysis

The modified surfaces can be subjected to X-ray diffraction (XRD) analysis to identify various phases present in the coatings. Diffraction spectrums are obtained by using suitable diffractometer of the desired type of target (e.g., Cu Kα radiations and nickel filter at current of 20 mA and 35 kV voltage). The specimens are

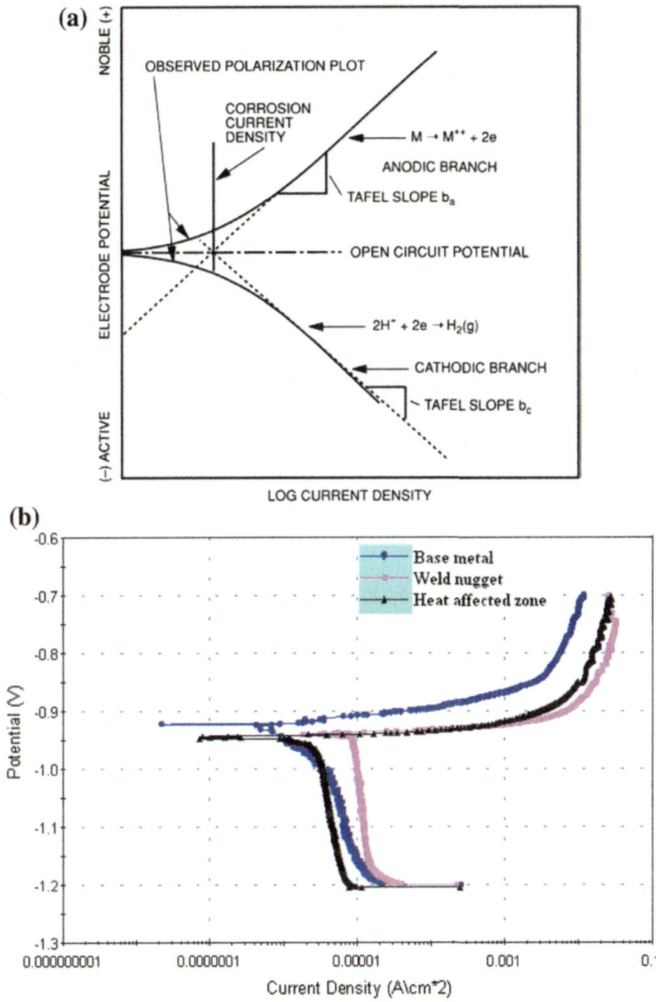

Fig. 7.37 Analysis corrosion data **a** schematic polarization curves and **b** actual polarization curves for Tafel analysis in 3.5% NaCl solution of different zone of FSW joints of AA 2024

scanned at a scanning speed of about 1.5 °/min through 20°–120° (diffraction angle "2θ"), and intensities are recorded at a chart speed of 1 cm/min with 1°/min as goniometer speed. Assuming height of the most prominent peak to be 100%, the relative intensities are calculated for all the peaks. The diffractometer interfaced with X-ray diffraction software provides interplanar spacing "*d*" and "2θ" values directly on the diffraction spectrum. Subsequently, these "*d*" and "2θ" values are used for the identification of various phases with the help of joint committee on powder diffraction files (JCPDF) or any other suitable data set.

7.9 Scanning Electron Microscopy (SEM)

Scanning electron microscopy (SEM) offers the benefit of studying the surfaces of different types of specimens like fracture surface and wear surfaces, e.g. as sprayed coatings, weld surfacing, etc., without any surface preparation. Scanning electron microscopy is also used to study the microstructure of the material and modified surfaces. A combination of high resolution and magnification as high as 100,000 permits easy analysis of fine microconstituents. SEM analysis of wear surface helps in investigating the mechanism of wear and extent of surface damage (Fig. 7.38). SEM images of worn surface are also used for studying the material removal mechanisms in abrasive and erosive wear conditions. Similarly, SEM analysis in fracture surfaces assists in establishing the mode of facture like dimple, cleavage, inter-granular, fatigue.

The microscopic examination of the fracture surface helps in identifying the operating micromechanism of the fracture and is usually carried out using devices like scanning electron microscope. Both transformation and scanning electron microscopes have different capabilities in terms of magnification and resolving power. The transmission electron microscope offers higher resolving power (up to 10 Å) and magnification (3×10^5) than the scanning electron microscope (up to 150 Å resolution and 1×10^5 magnifications). Specimens are usually coated with thin layer of gold of about 50 Å to make them electrical conductor and better reflection. SEM is more popular than transmission electron microscopy (TEM) because of two reasons related to sample preparation (a) sample preparation for TEM is very tedious and time consuming and (b) On contrary, no sample preparation is needed for SEM except that it should be small enough to be accommodated in vacuum chamber of SEM.

Depending upon the type of material and loading conditions, fracture surface may reveal variety of microscopic fracture mechanisms such as dimple fracture, cleavage fracture, inter-granular fracture, and fatigue fracture. The fracture based on macroscale deformation of the material (before fracture) can be classified as ductile fracture and brittle fracture. Among the four microscopic mechanisms of the fracture, in general, dimple fracture belong to ductile fracture, while the other three, namely cleavage, intergranular, and fatigue fractures correspond to brittle fracture. However, exception of these are also observed.

Dimple fracture is usually associated with extensive plastic deformation of materials prior to fracture, which is indicated by the presence of conical shape deep cavities in one of the fracture surfaces and corresponding conical shape protrusions in another fracture surface. The number, size, and depth of dimple suggest the extent of plastic deformation and load carrying capacity. Dimple fracture is considered as high energy fracture because it consumes lot of energy in causing plastic deformation prior to fracture. Fracture tough material of high load-carrying capacity and good ductility predominantly exhibit fine dimple fracture.

Cleavage fracture is associated with brittle fracture and characterized by the presence of typical river like pattern on the fracture surface that are formed because of intermittent growth of crack at different crystallographic plane leading to development of steps under the influence of external load. In cleavage fracture, cracks propagate through the grains that come across the cracks; conversely, it is a result of trans-granular fracture. Cleavage fracture is considered as low-energy fracture because it consumes little energy and usually offers low fracture toughness and limited deformation prior to fracture.

Intergranular fracture is also associated with brittle fracture and characterized by the presence of typical flat surfaced ball shape grain on the fracture surface formed by decohesion of grains under the influence of external load owing to the presence of some poor and brittle phases or compounds at grain boundary. In this type of fracture, cracks propagates mostly along the grain boundaries to cause the fracture; hence, it is termed as inter-granular fracture. Fractures occurring due to hydrogen-induced cracking, stress corrosion cracking, and sensitization of stainless steel, etc., fall under the category of intergranular fracture. Like cleavage fracture, intergranular fracture is also a low energy fracture with poor load-carrying capacity and limited ductility.

Fatigue fracture is mostly catastrophic and is generally characterized by three distinct regions on the fracture surface, which correspond to (a) fatigue fracture initiation site, (b) stable crack growth zone, and (c) sudden fracture zone. Fracture owing to the fatigue usually exhibits concentric circles with center at crack initiation site commonly termed as beach marks at low magnification and striations at high magnification. These features are developed during the second stage of fatigue fracture, that is, stable crack growth. According to the nature of material, the region corresponding to sudden fracture may show either dimple or cleavage fracture.

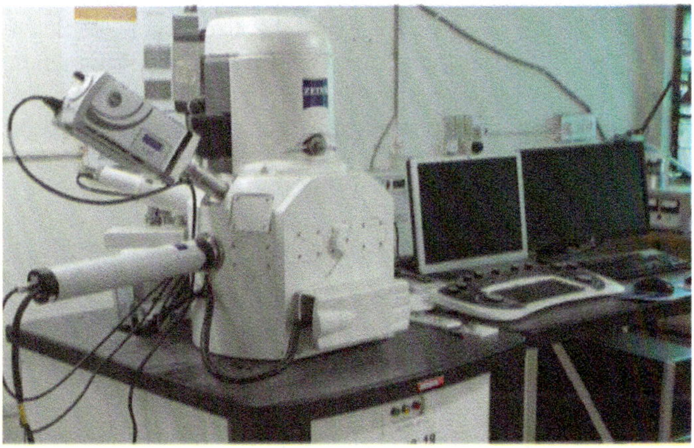

Fig. 7.38 Typical photograph of Fe-SEM system

7.10 Compositional Analysis

Mechanical, physical, and tribological properties of the surface are predominantly affected by its micro-constituents, which in turn depends upon its chemical composition. Besides thermal and mechanical history experienced during manufacturing Sometimes, efforts are made to modify the surface with a specific composition, but end up with different composition due to variety of reasons such as poor control of processing conditions, parameters such as temperature, time, and environment. For example, the surface modification processes involving melting, poor control of processes such as laser alloying, laser melting, weld surfacing can lead to excessive melting or lack of melting of substrate. The depth of melting affects the dilution percentage of alloying elements. Similarly, excessive longer exposure of melt at high temperature can lead to losses because of the reactions of elements with atmospheric gases. This loss of elements can occur in the form of oxides and nitrides and evaporation. Therefore, compositional analysis of modified surface becomes important for successful application of surface engineering. This can be performed using spectroscopy and energy dispersive X-ray analysis besides conventional wet technique.

7.11 Energy Dispersive X-Ray (EDAX) Analysis

Energy dispersive X-ray (EDAX) analysis is always performed in combination with SEM, and it provides an elemental analysis of microconstituents. It is performed in different ways, namely point analysis, line analysis, area analysis, and elemental mapping. Point analysis shows the elements present in a very small zone like a point using a beam of 100 nm to 1 μm diameter for analyzing a single point or multiple points as per choice. Line analysis gives the concentration of elements along the particular line selected between the two points of interest in the microstructure (Fig. 7.39). It is found useful to study the segregation or depletion of elements across the grain boundary or grain itself and variation in composition from surface of modified component to unaffected substrate. Area analysis provides the concentration of elements in a selected area in SEM image (Fig. 7.39). Elemental

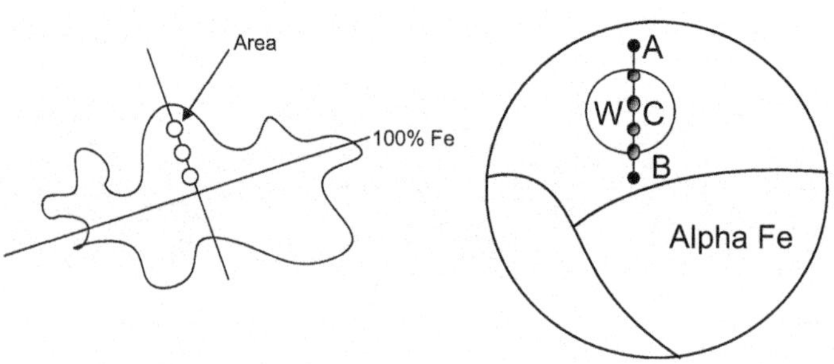

Fig. 7.39 Schematic of line scan and area analysis

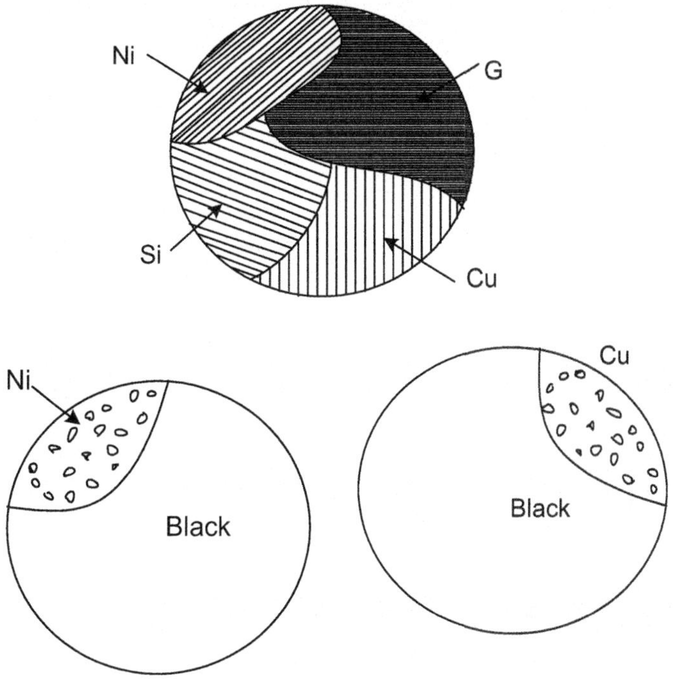

Fig. 7.40 Schematic of elemental maps showing distribution of elements

mapping is carried out for identifying the distribution of elements over different microconstituents in selected area of microstructure (Fig. 7.40).

Line scanning
See Figs. 7.39 and 7.40.

Spectroscopy
Electromagnetic radiation, ions, or electrons are directed toward the surface which is to be tested (Fig. 7.41). These penetrate the surface through few atomic thickness and then get reflected. The pattern and intensity of these reflections are recorded and compared with standard patterns of different elements for identifying the surface composition. This technique is used for elements of higher atomic weight (>5) because studying the patterns of these elements is easier.

7.12 Macroscopic Observation

Macroscopic observation of the modified performed in the range of 1–50 magnification with the help of lenses, stereoscope, and optical microscope (with external lighting) and more commonly with the scanning electron microscope. Plastic replicas

Fig. 7.41 Schematic of spectro-analysis approach

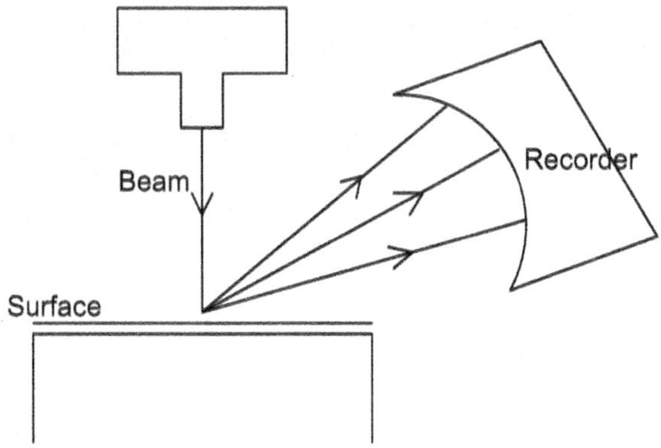

coated with gold layer of about 200 Å can also be used for macroscopic observation using SEM. A careful macroscopic examination can reveal important information about the location wherefrom cracks have initiated, direction of crack growth, and operational fracture mechanism.

7.13 Metallographic Examination

Metallographic examination is one of the most important tools available for metallurgical characterization of modified surface, as it helps:

- To assess the soundness and integrity of modified surface (for the presence of desirable and undesirable features such as unfavorable orientation of grains, porosity).
- To get idea about the suitability of composition.
- To study the effect of service and aging conditions such as decarburization, excessive grain growth, if any.
- To obtain the information about thickness/depth of modified zone.
- To determine the effect of environment such as corrosion, oxidation, work hardening.
- To identify the microstructural constituent contributing to crack nucleation and propagation, if any.

The sample of modified zone comprising substrate should be taken for metallographic studies for the analysis. Image analyzing software can be very useful to quantify the morphological characteristics of the microconstituents that can be related with performance. The morphological features include grain size, shape (aspect ratio, circularity, nodularity, form factor, shape factor, etc.), number of particles per unit area, relative amount of various phases, and their distribution (Fig. 7.42). Additionally, image analyzers can also help in measuring the geometrical dimensions of inclusion, cracks, and proportions of various micromechanisms

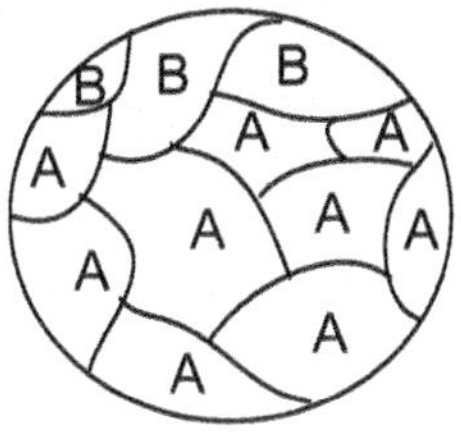

Fig. 7.42 Schematic of microconstituent

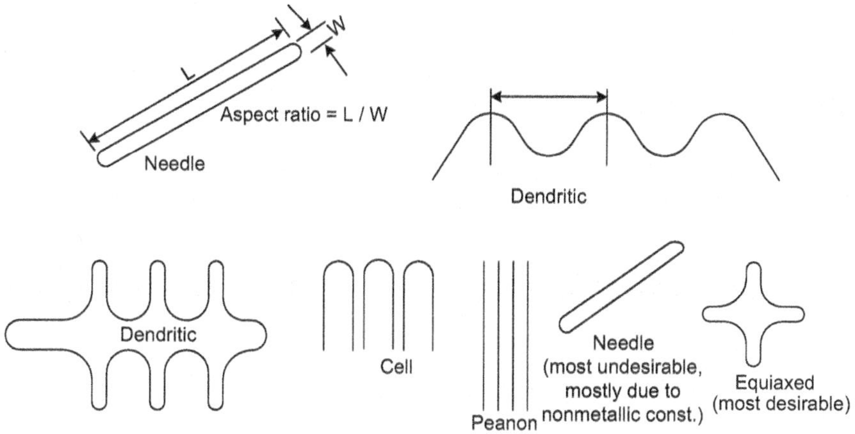

Fig. 7.43 Schematic of different shapes of microconstituents

(such as dimple, cleavage) present on the fracture surface. Microstructure of coating must be studied at three levels: (a) coating, (b) coating–substrate interface, and (c) substrate below interface. Microstructural features, namely grain shape such as spherical, nodular, dendritic, polyhedral, etc., size can be observed (Fig. 7.43). Further, grain size and arrangement, alignment and presence of inclusions, voids, cracks at surface or interface can be investigated.

When the coating is composed of entirely different composition from the substrate, elements from coating can diffuse into substrate at the interface, thereby causing a hard or soft band formation at the interface as well as degradation of coating and substrate, for example Ni–WC coating results in a black Crich band at coating substrate interface. Hardening at interface can take place when surface is heated to high temperature followed by rapid cooling. Softening at the interface can take place when the already hard surface is further heated during surface modification.

Typical procedure for metallographic studies

The bulk materials and engineered samples are cut by an abrasive wheel cutter. The cut samples are polished using the standard metallographic procedure. The samples are polished using different grades of water proof SiC abrasive papers (120, 220,

320, 400, 600, 800,1000 and 1200 grit size). Final polishing is carried out using cloth polishing wheel with 1 μm levigated alumina or iron oxide. Specimens are then washed in acetone and dried. For microstructural analysis, the samples are etched using suitable etchants. Micrographs of the etched samples are taken at different magnifications (starting from the low magnification (30–50×) to high magnification (200–500×), and these micrographs are used for microstructural analysis. The average grain size, percentage of various phases, distribution, shape of microconstituents in terms of shape factor, aspect ratio, circularity, etc., are determined by the image analysis of micrographs by using suitable image analysis software.

Questions for self-assessment

1. What is the significance of surface roughness on tribological life of the component?
2. Describe the methods used for measuring surface roughness.
3. How to characterize the thickness of the coating and explain the common methods for measuring thickness of the coating?
4. What is importance of coating–substrate bond strength in surface engineering?
5. Describe the method that can be used to characterize the bond strength of coating.
6. What are common nondestructive testing methods for assessing the soundness of the coating?
7. Describe the principle, application, and limitation of the following NDT methods:

   (a) Dye penetrant
   (b) Magnetic particle testing
   (c) Ultrasonic testing
   (d) Radiographic testing

8. What is the role of coupler in ultrasonic testing?
9. What is the significance of the hardness in surface engineering?
10. Explain the common principle used for hardness testing by different methods.
11. Describe methods, indenter, loading used for hardness by following methods.

    (a) Brinell
    (b) Rockwell
    (C) Vickers
    (d) Knoop

12. What for the microhardness testing is used? How to measure the subsurface microhardness to study the damage during service?
13. What is relevance of toughness of a material in engineering applications?

14. Describe methods, samples used for toughness test by the following methods.
    (a) Charpy
    (b) Izod
15. What is ductility to brittle transition temperature?
16. What are the parameters that need to be mentioned with toughness test?
17. What are the parameters that need to be mentioned with fatigue test report?
18. What is fracture toughness and describe the methods used for testing the same.
19. Explain the common methods and test parameters used for following adhesive wear test.
    (a) Unidirectional sliding wear
    (b) Reciprocating wear
20. Explain following method and test parameters used for abrasive wear test.
    (a) Pin on flat
    (b) Rubber wheel abrasive wear test
21. Explain following method and test parameters used for erosive wear test.
    (a) Solid particle erosion
    (b) Slurry erosion
    (c) Cavitation erosion
22. What is the importance of scanning electron microscopy, and XRD analysis in surface engineering?
23. Describe different approaches used for chemical analysis of microconstituents.

References

Chatterjee S., Pal T.K., Wear behaviour of hardfacing deposits on cast iron, Wear 255 (2003): 417–425.

Borle S.D., Gall I. Le, Mendez P.F., Primary Chromium Carbide Fraction control With Variable Polarity Saw, Welding Journal, (2015): 1-S.

Vespa P., Pinard P.T., Gauvin R., and Brochu M., Analysis of WC/Ni-Based Coatings Deposited, by Controlled Short-Circuit MIG Welding, JMEPEG (2012) 21:865–876.

Myers D., Metal cored wires, advantages and disadvantage, Welding Journal, 81 (2002): 39–42.

Lyttle, K. 1996. Metal cored wires: Where do they fit in your future? Welding Journal, 75 (10): 35–38.

Chai et al., basic metal cored electrode, US Patent, 5225661, (1993).

Deng Hongxia, Huiji Shi, Seiji Tsuruoka, Influence of coating thickness and temperature on mechanical properties of steel deposited with Co-based alloy hard facing coating, Surface & Coatings Technology 204 (2010): 3927–3934.

Corengia P., Ybarra G., Moina C., Cabo A., Broitman E., Microstructure and corrosion behaviour of DC-pulsed plasma nitride AISI 410 martensitic stainless steel at 500 C for 20 hrs, Surface & Coatings Technology 187 (2004): 63–69.

Wang X.H., Han F., Liu X.M., Qu S.Y. and Zou Z.D., (2008). Effect of molybdenum on the microstructure and wear resistance of Fe-based hardfacing coatings. Materials Science and Engineering A, 489(1–2): 193–200.

Dhakar B.M., Dwivedi D.K., Sharma S.P., Studies on remelting of tungsten carbide and rare earth modified nickel base alloy composite coating, Surface Engineering 28 (1): 73–80.

Arji R., Dwivedi D.K., Gupta S.R., Some studies on slurry erosion of flame sprayed Ni–Cr–Si–B coating, Industrial Lubrication and Tribology 61 (1): 4–10.

Sharma S.P., Dwivedi D.K., Jain P.K., Effect of La 2 O 3 addition on the microstructure, hardness and abrasive wear behavior of flame sprayed Ni based coatings, Wear 267 (5): 853–859.

Purkayastha S., Dwivedi D.K., Abrasive and Erosive Wear Performance of Rare Earth Oxide Doped Ni/WC Coatings, Journal of Tribology 136 (2014): 011602.

Harsha S., Dwivedi D.K., Agarwal A., Influence of CrC addition in Ni–Cr–Si–B flame sprayed coatings on microstructure, microhardness and wear behaviour, The International Journal of Advanced Manufacturing Technology 38 (1–2): 93–101.

Dwivedi D.K., Abrasive wear behavior of iron base hard surfacing alloy coatings developed by welding, Surface Engineering, Vol. 20, No. 2, (2004), 87–92.

Dwivedi D.K., Microstructure and abrasive wear behavior of iron base hardfacing developed by SMA welding, Material Science and Technology, 20, 10, (2004): 1326–1330.

D. K. Dwivedi, *Surface Engineering*,
https://doi.org/10.1007/978-81-322-3779-2

Sharma Tarun, Maria Saurabh, Dwivedi D.K., Abrasive Wear Behaviour of Fe-30Cr-3.6C Overlays Deposited on Mild Steel, ISIJ International, 45, 9, (2005): 1322–25.

Harsha S., Dwivedi D.K., Agarwal A., Investigations on the abrasive wear behavior of flame sprayed Ni–Cr–Co–Si alloy coating deposited on mild steel substrate, ISIJ International, 46, (2006), 10, 1473–78.

Yadav A.K., Arora N., Dwivedi D.K., On Microstructure, Hardness and Wear Behaviour of Flame Sprayed Co Base Alloy Coating Deposited on Mild Steel, Surface Engineering, 22, 5 (2006): 331–336.

Shah K.B., Kumar S., Dwivedi D.K., Abrasive wear behavior of Fe–Cr–C overlays, Paton Welding Journal, 11, (2006): 22–27.

Harsha S., Dwivedi D.K., Agarwal. A., Influence of WC Addition in Co–Cr–W–Ni–C Flame Sprayed Coatings on Microstructure, Micro hardness and Wear Behaviour Surface and Coatings Technology, Surface and Coating Technologies, 201 (2007): 5766–5775.

Harsha S., Dwivedi D.K., Some studies on Microstructure, Hardness and Abrasive Wear Behaviour of Flame Sprayed Co base alloy Coating, Surface Engineering, 23 4, 2007, 261.

Arji Rambabu, Dwivedi D.K., Gupta S.R., Sand Slurry Erosive Wear of Thermal Sprayed Coating of Stellite, Surface Engineering, 2007, 23, 5, 391–97.

Harsha S., Dwivedi D.K., Agarwal A., Performance of Flame Sprayed Ni-WC Coating under Abrasive Wear Conditions, Materials Engineering and Performance, 17, 1 (Feb 2008) pp 104–109.

Sharma S.P., Dwivedi D.K., Jain P.K., Microstructure, hardness and abrasive wear behavior of continuously compacted thermal sprayed Ni base alloy powder coatings in different conditions, International Journal of Surface Science and Engineering, 2, 3–4, (Sep 2008) 240–251.

Sharma S.P., Dwivedi D.K., Jain P.K., Effect of CeO_2 addition on the microstructure, hardness and abrasive wear behavior of flame sprayed Ni based coatings, Proceedings of the I MECH E Part J Journal of Engineering Tribology, 222, 7, (Nov 2008), 925–933.

Rambabu Arji, D. K. Dwivedi, S. R. Gupta, Some Studies on Slurry Erosion of Flame Sprayed Ni–Cr–Si–B Coating, Industrial Tribology and Lubrication, 61, 1 (Jan 2009), 4–10.

Jagannadham D., Dwivedi Dheerendra Kumar; Ghosh P.K., Weld Surfacing of Aluminium alloy by Ni–Cr Base Alloys, Proceedings of the Institution of Mechanical Engineers, Part L, Journal of Materials: Design and Applications, Vol 233, (Aug 2009) 117–129.

Sharma S., Dwivedi D.K, Jain P.K, Study of mechanical and metallurgical characteristics of flame sprayed NiCrBSi as sprayed and continuous compacted, Proceedings of the I MECH E Part J Journal of Engineering Tribology, Vol 224, (Jan 2010) 107–114.

Dwivedi D.K., Adhesive Wear Behaviour of Cast Al–Si Base Alloys: Overview, Materials and Design, 31 (Feb 2010), 2517–2531.

Sharma S.P., Dwivedi D.K., Jain P.K., Abrasive Wear Study of Flame Sprayed Co-Based Composite Coatings, Journal of Tribology and Surface Engineering, Vol. 2 (2011), 41–48.

Kiran S.R., Gandhi B.K., Dwivedi D.K., Paul C.P. and Kukreja L.M., Erosive Wear Behaviour of Laser Clad Surfaces of Ni and Co Based Alloys, Journal of Tribology and Surface Engineering, Vol. 2 (2011), 33–40.

Dhakar B.M., Dwivedi D.K., Sharma S.P., Studies on re-melting of tungsten carbide and rare earth (RE) modified Nickel base alloy composite coating, Surface engineering, Vol. 28 (Jan 2012), No. 1, 73–80.

Sharma S.P., Dwivedi D.K., Jain P.K., Study and model development of erosive wear by RSM, Proceedings of the I MECH E Part J Journal of Engineering Tribology, 226 (June 2012), 57–70.

Surajit Purakayastha, D K Dwivedi, Abrasive and erosive wear performance of rare earth oxide doped Ni/WC coatings, ASME Journal of Tribology, Vol. 136 (2014), 602–609.

Paul C.P., Gandhi B.K., Bhargav R., Dwivedi D.K., and Kukreja L.M., Cobalt Free Laser Cladding on AISI type 316L Stainless Steel for Improved Cavitation and Slurry Erosion Wear Behavior, Journal of Materials Engineering & Performance, 23, (Sep 2014), 4463–4471.

Singla Y.K., Dwivedi D.K., Arora N., On the modeling of dry sliding adhesive wear parameters of vanadium additive iron-based alloys at elevated temperatures, Surface and Coatings Technology, 284, 223–233.

Thapliyal Shivraman, Dwivedi Dheerendra Kumar, Study of the effect of friction stir processing of the sliding wear behavior of Cast NiAl Bronze: A statistical analysis, Tribology International, 97 May 2016) 124–135.

Thapliyal Shivraman, Dwivedi Dheerendra Kumar, Microstructure evolution and tribological behavior of the solid lubricant based surface composite of cast Nickel Aluminum Bronze developed by friction stir processing, Journal of Materials Processing Technology, 238, 30–38.

Singla Y.K., Arora N., Dwivedi D.K., Dry Sliding Adhesive Wear Characteristics of Fe-based Hard facing Alloys with different CeO_2 Additives–A Statistical Analysis, Tribology International, 105, 229–240.

Verma Sandhya, Dwivedi D.K., Chandra Ramesh, Cavitation Erosion Behavior of Nitrogen Ion Implanted 13Cr4Ni Steel, International journal of Trans. Of Indian Institute of Metals, (accepted).

Dwivedi D.K., Transitions in wear-friction behavior under dry sliding conditions of cast Al-Si base alloys Institution of Engineers (India), Vol. 82, (2001), 69–74.

Saini N., Dwivedi D.K., Jain P.K., Singh H., Surface Modification of Cast Al-17% Si Alloys Using Friction Stir Processing, Procedia Engineering 100, (2015) 1522–1531.

Dwivedi D.K., Maintenance of worn out surfaces: Hard surfacing and Thermal Spraying, in proceedings of national seminar on Maintenance and condition monitoring, at Govt. Engineering College, Thrissur, w. e. f. Feb.14, (1998), 174–181.h.

Dwivedi, D.K., Studies on wear behavior of iron base hard surfacing alloy coatings developed by different welding processes proceedings of National Convention of Mechanical Engineers on Research and Development in Mechanical Engineering, at MIED, UOR, Roorkee, w. e. f. Sep. 29–30, (2000), 84–87.

Dwivedi D.K., Influence of Heat Input on Microstructure and Wear Behaviour of Fe Base Overlays Deposited on C Steel. Presented at NMD-ATM 2005 held on Nov. 15–16, (2005) at IIT Madras, Chennai.

Harsha S., Dwivedi D.K., Agarwal A., Some studies on wear behavior of flame sprayed nickel and cobalt base alloy coatings deposited on steel, AIMTDR, 2006 w. e. f. Dec-21–23, 2006.

Dwivedi D.K., Combating Wear of Components in Sugar Industries, present and published in proceedings of International workshop on Advances in Surfacing Technologies Chennai (2007) w. e. f. June 29–30, 2007.

Sharma S.P., Dwivedi D.K., Jain P.K., Effect of La_2O_3 addition on the microstructure, hardness and abrasive wear behavior of flame sprayed Ni based coatings, 17th International Conference held at Las Vegas, USA, w.e.f. April 19–23, 2009.

Sharma S.P., Dwivedi D.K., Jain P.K., Abrasive wear response surface model of HVOF sprayed NiCr–WC–Co composite coatings, Proceedings International Conference on Surface Modification Technologies (SMT 23) held at Chennai, w.e.f. Nov 2–5, 2009.

Dhakar B.M., Dwivedi D.K., Sharma Satpal, Influence of TIG re-melting and RE (La_2O_3) addition on microstructure, hardness and wear of Ni–WC composite coating, Presented and published in proceedings of symposia on surface coatings and films in TMS 2012 to held at Orlando, USA w.e.f. March 11–15, 2012.

Purkayastha S., Dwivedi Dheerendra Kumar, Effect of CeO_2 on the friction and sliding wear performance of Ni/WC coatings, presented and published in proceedings of International conference on Advances in Electrical and Mechanical Engineering, Phuket, Thailand, Dec 18–19, 2012.

Sinhmar Sunil, Dwivedi D.K., Pancholi Vivek, Friction stir processing of AA 7039 alloy, presented and published in proceeding of International conference on Production and Mechanical Engineering, Bangkok, Dec 30–31, 2014.

Thaplial Shivraman, Dwivedi Dheerendra K., Microstructural and mechanical characterization of multipass friction stir processed cast nickel aluminium bronze (C95500) alloy, presented in International conference on tribology and interface engineering, held at Kuala Lumpur during June 11–12, 2015.

Arji R., Dwivedi Dheerendra Kumar, Erosive Wear Behaviour of Ni–WC Coatings Developed by Flame Spraying, International Conference on Innovation in Engineering, held at Mauritius, during March 20–21, 2016.

S. M. Sharif, Laser assisted surface modification of ferrous system, Lecture in continuing education program on combating wear held during 29-5-2007 to 02-6-2007.

Singla Y.K., Development of Rare earth Modified hardfacing fluxes for enhanced wear resistance (2016), IIT Roorkee, India.

Sharma S., Abrasive and erosive wear of thermal sprayed coatings (2009), IIT Roorkee, India.

Kiran S. Raja, Development and characterization of Cavitation resistant non-cobalt base laser cladding, M. Tech thesis (2009), MIED IIT Roorkee.

Arji Rambabu, Investigation on erosive wear behavior of Ni–Cr, Co–WC coating deposited by detonation gun on stainless steel substrate, M. Tech thesis (2006), MIED IIT Roorkee.

Dhakar B.M., Wear studies on thermal sprayed coatings on steel. M. Tech thesis (2010), MIED IIT Roorkee.

Purkastha Surajit, The mechanical and tribological performance thermal sprayed component, M. Tech thesis (2010), MIED IIT Roorkee.

Boronizing: http://en.alper.com.tr/hizmetler/isil-islem/boronizing.html.

Radiography: www.ndt.kalkars.com/radiographic-test-film-interpretation.html.

Index

D. K. Dwivedi, *Surface Engineering*,
https://doi.org/10.1007/978-81-322-3779-2

The manufacturer's authorised representative in the EU is Springer Nature Customer Service Centre GmbH, Europaplatz 3, 69115 Heidelberg, Germany. If you have any concerns regarding our products, please contact ProductSafety@springernature.com

Printed and bound by CPI Group (UK) Ltd, Croydon, CR0 4YY

29/07/2026

02180519-0001